Additional Praise for *Between the Sea and the Sky*

"An excellent study and analysis of 'lived religion' that integrates the dangerous and risky work of a fishing community of southwest India with their ritualistic Christian faith, centered around the protective warrior saints and their own neithal/Indian spirituality that relates them to and protects them from the challenges of the sea. A very good example of enculturated and contextual theology!"

– **Michael Amaladoss**, SJ, Indian theologian and author

"To those habituated to theologies from the land, here is a highly fascinating and extremely illuminating volume of theology from the sea. It dwells in the cosmology, life world, and everyday practices of a group of Christian fisherfolk whose vision, values, moods, religious experiences, and motivations are drenched in the vagaries of the nurturing and, at the same time, menacing oceans."

– **Felix Wilfred**, founder and director,
Asian Centre for Cross-Cultural Studies, Chennai

"The seaside communities of southern India are home to intensely devoted Catholics who trace their Christian roots to Saint Francis Xavier. The lives of Catholic Mukkuvar fishermen there are oriented to the sea—a bounteous, yet fickle, sometimes dangerous source of well-being that gives rise to its own unique cosmology. P. T. Mathew has a keen eye not only for how fisherfolk have been shaped by the sea but also how they have shaped and interpreted Christian faith. This is a great contribution to our understanding of Indian Catholic life."

– **Thomas M. Landy**,
College of the Holy Cross

"Lived religion is in the blood and veins of the Catholic fisherfolk of Kerala. It forms their identities, instils them with infinite courage, and provides patience and fortitude to accept the unpredictable outcomes of fishing. Mathew's erudite rendering of this complex socioreligious and cultural reality stands out for its deep and nuanced understanding of the fluid and boundless, yet coherent worldview of the fishers, which straddles the sea and sky on every fishing trip."

– **John Kurien**, former vice-chair,
FAO/UN Advisory Committee for Fisheries Research;
founding member, International Collective in Support of Fishworkers

"P. T. Mathew's splendid new book tells the story of the indigenous fisherfolk of Thiruvananthapuram coast of Kerala. Informed by critical ethnographic and historical scholarship, *Between the Sea and the Sky* offers a vivid portrait of a people's resilience rooted in their lived religion—a rich interweaving of Indic/neithal religiosity and Iberian Catholicism. In an age of unrelenting globalization, Mathew's study of neithal Catholicism reminds us that wisdom is lived in stories that, like the people themselves, remain unique and irreplaceable."

– **William O'Neill**, SJ,
University of San Francisco

"The book is a fabulous model for doing theology in context. The anthropological, ethnographical, and sociological analysis of the worldview, belief system, ritual practices, and folk traditions of the fisherfolk makes it a unique specimen of scientific research. This volume is highly informative regarding the constituents of lived neithal religiosity, a topic hardly explored by the intellectual world. The striking portrait of fishermen's way of overcoming the tsunami of 2004 and Ockhi cyclone of 2017 and of running rescue operations during the Kerala floods of 2018 with their traditional wisdom and daring faith is, in its content as well as recitation, a brilliant pattern of narrative theology."

– **Vincent Kundukulam**, president,
Indian Theological Association

"*Between the Sea and the Sky* is a wonderful work on the life of a fishing community in Kerala in South India, who struggle hard to survive amidst the challenges posed by both nature and the prevailing social conditions. P. T. Mathew shows us how the lived religion of the fisherfolk who spend most of their time between the sea and the sky serves as a vital source of energy to face their life challenges. This well-researched work is a resourceful book for those interested in understanding the organic relationship between religion, sea, and the fisherfolk."

– **R. Sooryamoorthy**,
University of KwaZulu-Natal, South Africa

"In an excellently lucid articulation, P. T. Mathew captures the rich experience of neithal Catholicism, the 'lived religion' of the fisherfolk of a South Indian region. In his descriptive and interpretative gaze, the subaltern self of a people living in a periphery comes to life in a poignant way through a religious idiom. It is a must-read for students of religious studies, theology, sociology of religion, ethnography of religion, and the like."

– **Gnana Patrick**,
University of Madras

BETWEEN THE SEA AND THE SKY

LIVED RELIGION ON THE SEASHORE

P. T. MATHEW, SJ

FORTRESS PRESS
MINNEAPOLIS

BETWEEN THE SEA AND THE SKY
Lived Religion on the Seashore

Cover image: Shifaaz shamoon on Unsplash

Cover design: Joe Reinke

Paperback ISBN: 978-1-5064-5199-2

eBook ISBN: 978-1-5064-5200-5

CONTENTS

List of Figures

List of Images

A Note on Orthography

Malayalam, Tamil, and Sanskrit terms in the text are italicized. The local mode of pronunciation has been adopted in the transliteration as far as possible. Diacritical marks follow the usage in the *Malayalam Lexicon,* published by the University of Kerala (Vol. I, Thiruvananthapuram, 1965). Place names and proper names that are widely used have been retained without diacritical marks.

PREFACE

The sea and the seashore have always been places of fascination and wonderment. The ocean has treasures that are valuable to life on earth yet poorly comprehended. This paradox is equally true of people who live by the ocean and depend on its bounty for their survival. Theirs is a world that is unique, as it is regulated more by what happens in the sea rather than on the landmass. This fascination led me to undertake a preliminary study of the world and the worldview of the coastal people two decades ago. The main purpose was to explore their world in order to understand the factors that enable anyone from the community to venture into the roaring sea on a tiny canoe with a daring spirit that is rarely seen elsewhere. That study, though confined to a coastal hamlet in the southern Indian State of Kerala, gave rich insights into the treasure house of wisdom and know-how that the fisherfolk have inherited through generations. It also unveiled the role of their religious worldview in bolstering their everyday life and occupation. The unpredictability of the ocean and the risk-ridden occupation necessitate an anchoring in a sound belief system and ritual practices.

The tsunami of 2004 gave an opportunity to the fisherfolk of the Kerala coast to take stock of their strengths and weaknesses in a crisis situation. The tsunami hit many countries on the rim of the Indian Ocean and devastated coastal areas in the southern States of India with thousands of people losing their lives and many more losing their means of livelihood. The memory of the tsunami was still alive when another disaster in the form of Cyclone Ockhi hit the Kerala coast in 2017, swallowing up hundreds of fishermen who were out in the sea fishing. In spite of such shattering experiences, the fishermen of the Thiruvananthapuram coast proved to be very resilient to danger and disaster and capable of carrying on with optimism. The same daring spirit was evident during the deadly floods in the State in August 2018, when the marine fishermen became saviors to thousands of flood victims. The Chief Minister of the State hailed these fishermen as the *State's Army*. That was a moment for the Kerala society to discover the strength and vitality of the marine fisherfolk, and also for the coastal people to grow in an awareness of their own resourcefulness. This resourcefulness and the never die hope of the marine fisherfolk was a major reason for me to revisit them for a fresh inquiry.

My earlier study that was done as part of a PhD program in the University of Madras was published in 2001 under the title, *We Dare the Waters–The World and Worldview of Mukkuvar*. The present book is a revised and updated version of it. I am grateful to the department of Christian studies, University of Madras, for the kind permission to bring out a revised edition of the book. The theoretical framework has been modified to incorporate the notion of lived religion paying attention to the changes that have taken place during the past two decades. The assumption is that religious worldviews are ever evolving in response to the changes taking place in the life context of a people. One of the chapters of the present book was earlier published as an article titled, "Mukkuva Cosmology," in the *Loyola Journal of Social Sciences*, Vol. 12, No. 1 (January–July, 1998, 39–59). I express my thanks to the Journal for the consent to use the article in the present book.

I am indebted to a number of people for helping me make this revised and updated version of the book a reality. The Kerala Jesuit Province relieved me from many responsibilities and gave me full support for the timely completion of this work. Professor Felix Wilfred, my Guru, continues to be an encouraging mentor. Professor R. Sooryamoorthy, Dean and Head of School of Sociology, Howard College Campus University of Kwazulu, Natal, South Africa, graciously found time to go through the manuscript and give me valuable suggestions. I am indebted to him. Many friends from the coastal belt accompanied me all through the study and contributed richly with their insights. I must specially mention A. Antony from Vizhinjam, Michael from Poonthura, A.J. Vijayan (Coastal Watch), Robert Panipilla (Friends of Marine Life), Fr Justin Judin (vicar of Vizhinjam church), Fr James Culas, Fr Joe Thayil, S.J., Dr E.J. Thomas, S.J., and Dr Iris Coeleo for sharing their knowledge and insights at various stages of the book's preparation. Alban, Raymond, and Lawrence were survivors of Cyclone Ockhi who willingly shared their near-death experience in mid-sea. Focus group discussions with active fishermen and activists from the coastal community were rewarding. Discussion with the MSW (Master of Social Work) students of Loyola College of Social Sciences, Thiruvananthapuram was also rewarding since they had actively involved in relief work among the Ockhi-affected people. I am grateful to Fortress Press for accepting to publish the revised version. I believe that this book will be useful to researchers, social activists, policymakers, religious leaders, and theologians who are laboring to make life on the seashore more wholesome.

CHAPTER 1

LIVED RELIGION IN TIMES OF CRISIS

Cyclone Ockhi and a New Realization

It is in times of crisis or disaster that people come to realize the power or powerlessness of the belief system they rely on. Cyclone Ockhi in 2017 was such an occasion for the fisherfolk of the Kerala coast. The very severe cyclonic storm with a wind speed of up to 155 km/hour hit the southern tip of the Indian peninsula on November 30, 2017. It claimed over 300 lives, either dead or confirmed missing.[1] They were all fishermen from the coastal belt of the two southernmost districts of Kerala and Tamil Nadu. All of them were engaged in fishing in their traditional boats or canoes oblivious to the impending disaster as the cyclone warning system was late in sending the danger signals. The depression that originated in the Bay of Bengal hit the eastern coast of Sri Lanka as a deep depression and intensified as a cyclone and partially hit the southern states, and moved away from the coast northwards, finally making its landfall on the Gujarat coast as a weak depression. Among the most affected centers in Kerala were the two coastal hamlets of Poonthura and Vizhinjam in Thiruvananthapuram district. Though the Indian Meteorological Department claimed that the cyclone did not make its landfall on the Kerala coast, the harm was already done; hundreds of fishermen who were already out in the sea were caught unaware. Global weather monitoring agencies had identified the formation of Ockhi but the official machinery failed to provide timely warning and to take precautionary measures. The anger of the coastal people was primarily directed against the State administration and the Indian Meteorological Department (IMD) for their negligence. News channels and newspapers

[1] The government stated in the Parliament on December 27, 2017, that a total of 661 fishermen, 261 from Kerala and 400 from Tamil Nadu, were still missing. Thiruvananthapuram Catholic Archdiocese, based on parish-wise data collection, found that 288 fishermen were either dead or missing within its territory. See *Samanvaya—Thiruvananthapuram Archdiocesan Bulletin*, June 2018, 25.

carried features depicting the misery on the shore with hundreds of families losing their dear ones.

Usually, cyclonic storms are very rare on the western side of the Indian peninsula unlike on the eastern side, and because of that the cyclone disaster on the Kerala coast became especially noteworthy. Cyclonic storms are frequent in the Bay of Bengal and periodically hit the eastern coast bringing calamity to the eastern states of Orissa, Andhra Pradesh, and Tamil Nadu. Bangladesh, the neighboring country, is also not spared. Since cyclones are an annual feature on the eastern coast, these states are better prepared to face the impact. Kerala, on the other hand, never expected such a calamity, and so remained unprepared. Studies show that the constitution of the Arabian Sea has been changing drastically in recent times. The surface temperature has increased significantly from an average temperature of 28°C to the present 31°C with an accompanying increase in relative humidity.[2] This could be one reason for the frequent occurrence of cyclones. Climate change is a fact that we have to reckon with.

The aftermath of the cyclone witnessed two streams of intense debate in the sociopolitical and religious spheres of Kerala. The first referred to the apparent failure of the weather monitoring agencies in the timely detection of the cyclone and its proper communication to those in the danger zone. The role of the government both in the State and at the Centre became part of this discussion. The second stream centered on the promptness and efficiency of disaster management and rescue operations that brought laurels to the State government. The role of the Church became significant since the vast majority of the affected fishermen belonged to the Latin Catholic Archdiocese of Trivandrum.The diocesan authorities state that 95 percent of the victims belonged to this diocese. The church was active in the field and in the media from the very beginning of the disaster, and engaged in rescue operations and organized prayer meetings for the safe return of those missing. Parish groups gathered details of those still missing or confirmed dead and displayed their photos near churches. Temporary shelters were set up to keep vigil and to engage in prayer. Volunteers and non-governmental organizations became active with food and other support. The channels and newspapers carried photographs of the fisherfolk in agony. The Archbishop and the leading priests of the diocese appeared in the news media speaking on behalf of their "flock." The deep connection between cyclone Ockhi, the marine fishermen, and the Latin Catholic Church was very evident.

The debate continued unabated touching upon improving the cyclone warning system, disaster preparedness, and timely compensation

2 A. Rajagopal Kamath, "Fear over the Sea", *Deccan Chronicle*, December 17, 2017.

to the affected families. In the meantime, the fishermen were largely left to fend for themselves. They had relied on the power of their traditional maritime wisdom, known as *kaṇiyam*, but that too seemed to fail them. "Never in my life as a fisherman have I seen a cyclone of this intensity," said a fifty-year-old fisherman from Poonthura. An early warning system and disaster preparedness are essential, but what concerns us in this study is the preparedness of the fisherfolk to face such disasters and the manner of their coping with these. The vital link between the risk-ridden occupation of fishing and the religiocultural ethos of the fisherfolk become the focal point here. Our assumption is that the fisherfolk display a remarkable degree of daring spirit even in disaster and amazing resilience afterwards.

The traditional method of fishing involves much risk since the behavior of the ocean is not always predictable and since traditional fishermen have limited access to life-saving equipments. Seasons of rough weather, tidal waves, cyclones, and tsunamis place fishermen in a zone of constant danger. Modern gadgets like compass and GPS (Global Positioning System) are not widely in use among them. Accidents occur especially during the monsoon season (June–August). Many lose their lives. The number of the dead or missing becomes newsworthy only in times of a cyclone or tsunami. Yet the resilience of the coastal people enables them to pick up the threads and move on. Orissa and Bangladesh can give us good insights. Cyclones are a regular feature on the eastern coast, and after the deadly Super Cyclone of 1999 that killed more than 10,000 people, Orissa was well prepared when Cyclone Phailin hit its coast in 2013.[3] Disaster studies in Bangladesh point to the importance and value of the local people, especially their perceptions and skill in coping with disaster for the success of any hazard mitigation program.[4] The attitude and behavior of the coastal people in relation to the ocean, their belief system and the goals they set for themselves, and the inherited skills they possess—all become handy in times of disaster. If the people of Europe had to struggle with questions of theodicy after the Great Lisbon earthquake of 1755, the outlook of the fisherfolk of the Kerala coast stands in clear contrast.

3 See Nibedita S. Ray-Bennett, "Learning from Deaths in Disasters: The Case of Odisha, India", University of Leicester (June 9, 2016), accessed August 19, 2018, http://www.mei.edu/content/map/learning-deaths-disasters-case-odishaindia.

4 Cf. E. Alam and A.E. Collins, "Cyclone Disaster Vulnerability and Response Experiences in Coastal Bangladesh" in *Disasters*, 34 (2010): 931–954, accessed August 14, 2018, http://dx.doi.org/10.1111/j.1467-7717.2010.01176.x.

The Lived Religion on the Seashore

Religious phenomenon constitutes a multifarious reality that eludes the grip of any single discipline in isolation. The social sciences are essential tools in the attempt to understand religion as lived as it falls within the empirical realm. Sociology of Religion shows clearly that religious phenomena are essentially social facts, linked to other social facts. At the same time, it admits to a non-empirical, transcendent dimension in religion. The area of human motivation and life orientation in suffering or anguish, amidst frustration or chaos, is a point to note. The social sciences cannot exhaust the multifarious complexity of religion or of human life. An approach that incorporates effectively the discipline of theology with the social sciences alone can effectively deal with this complexity and contribute to a proper understanding of religion as lived by a people.

A historical overview of sociological studies of religious phenomena reveals three distinct perspectives as Michael Hill helps us to understand:[5]

a) Religion considered as an autonomous, self-regulating aspect of human experience that is beyond the influence of material and social environment—a purely transcendental phenomenon. This is a purely de-historicized theological perspective and would place religion outside the scope of sociology. The prescriptive definition of religion by Paul Tillich is an example: "God is the answer to the question implied in man's finitude; he is the name for that which concerns man ultimately."[6]

b) Religion understood and explained purely in terms of economic and social environment. It excludes, by implication, those understandings of religion that imply a belief in a transcendent divinity (e.g., Karl Marx, Emile Durkheim, J. Milton Yinger, and others). This approach is reductionist and finally explains religion away.

c) Religion recognized as a social phenomenon, where religious beliefs and experiences are treated as social facts, yet admitting its transcendental dimension as central to it. This may be called *sociotheological* perspective. Robert Bellah, Thomas Luckmann, and Peter Berger are exponents of this outlook. The present study may be considered socio-theological in this sense.

Religion plays a vital role in the daily life and occupation of marine fisherfolk, irrespective of the religious tradition they belong to. What

---

5 Adapted from Michael Hill, *A Sociology of Religion* (London: Heinemann, 1980), 4–17.

6 Paul Tillich, *Systematic Theology* (Chicago: The University of Chicago Press, 1971), 211.

usually comes under scrutiny in any study of religion is its rational articulation comprising its scripture, doctrinal codifications, ethical precepts, and institutional structure. Robert Merton makes a distinction between the *manifest* and the *latent* modes of religion.[7] He explains the m*anifest mode* as referring to the socially recognized and sanctioned elements of religion that exist within conventional and official structures and norms. The *latent mode*, on the other hand, refers to those elements that are not officially recognized and approved, and hence operate outside the official structures and norms.[8] Lesser visibility would go with the latter. In this study, the concept of "lived Catholicism" is used synonymously with the "latent mode of Catholicism" as discussed above. Authors like Meredith McGuire use the concept of lived religion as the ethnographic framework for understanding beliefs and practices of the faithful in their everyday life, taking into account the nonofficial and the non-dogmatized elements in the practice of religion.[9] Deep down, it is lived religion that would reveal the world of subjective understanding of a faith community and explain its actual universe of meaning.

One distinguishing mark of lived religion is related to its distinct approach to the religious boundary. The boundary is a notion that is usually employed to designate the separating line between two realms or regions, whether conceptual or geopolitical. A fence, a wall, a river, a mountain, or a forest may become a physical boundary line. The notion comes to us primarily as referring to a nation-state: as administrative units having a monopoly of power over a territory with demarcated boundaries legitimizing its rule by laws it enacts. The same concept is being applied to social segments like caste groups, ethnic or tribal collectives, and religious communities. It is in this conceptual sense that we refer to the religious boundary. The boundary can be understood either as a line of separation (visible boundary) or as a region of interaction (invisible boundary); boundary in the former sense "encloses," while in the latter sense it "opens up." The enclosing provides a visible identity separating it from other contending religious groupings. Simultaneously, it stands in contrast to what is outside and what is unfamiliar. The unfamiliar has a different horizon of meaning, a different worldview. The boundary does demarcate and clarify, yet always remains perplexing and mysterious. A boundary reveals both our limits and the inexhaustible mystery that lies beyond. Understood thus, a boundary is an ambivalent concept. Conflict

7 Robert K. Merton, *Social Theory and Social Structure*, rev. ed. (New Delhi: Amerind Publishing,1972), chapter 3.

8 Merton, *Social Theory*, chapter 3.

9 Meredith McGuire, *Lived Religion: Faith and Practice in Everyday Life* (Oxford: Oxford University Press, 2008), 12.

characterizes a boundary when it is seen purely as a line of separation. The more rigid a boundary is, the more strange what remains outside of it appears. The official Catholic tradition, rooted in its Semitic worldview, looks at boundary primarily as a line of separation, while lived religion tends to view it as a region of interaction. Lived religion in the present study refers to the Catholicism as lived by the fishing communities along the Kerala coast in the southern corner of the Indian peninsula.

Two decades have passed since I had undertaken a preliminary study of the world and worldview of the coastal people of Thiruvananthapuram coast. Much has happened on the seashore during this period. The present book is an attempt to revisit the coastal community in the changed situation in order to see how the fisherfolk have refashioned their religious worldview in sustaining themselves in good times and in hard times. Religious worldviews are ever evolving in response to the changes taking place in the life context of a people. This continuity and discontinuity with the coastal situation of the 1990s characterized the revised version. The theoretical perspective has been modified to incorporate the changed situation adequately.

The fisherfolk of Kerala belong to any one of the three dominant religious segments in the State, viz., Hindu, Muslim, and Christian. Over the years, each group has developed a subculture of its own within a broad coastal culture. This study does not aim at a general survey of religious phenomena among the three groups nor is it a comparative study of the three as both would imply superficial generalizations. Religions are incomparable as each religion has its own universe of meaning. So, we lack a common yardstick with which religions can be compared. To its adherents, each religion would appear a perfect and complete whole. To get a valid insight into the religious phenomena among the fisherfolk, it is necessary to base our inquiry on a specific faith community that at the same time has the identity as a specific cultural group. The choice of the Catholic fishing community on the Thiruvananthapuram coast in Kerala is prompted by two reasons. First, the Catholic community constitutes the major fishing segment in the district. Second, the fishing harbor at Vizhinjam on this coast has been a nerve center of fishing activity in the southern Kerala with a natural bay and a vibrant fishing community. My familiarity with the locale as well as the coastal community for many years also contributed to the choice.

The objectives of this study can be stated as follows:

a. How does lived religion become the source of energy and courage for the marine fisherfolk in their risk-ridden occupation?

b. How can we grasp the components and dynamics of the lived religion of the coastal Catholic fisherfolk? What role does Catholic faith play in this?

c. What are the critical elements missing in their lived Catholicism that need to be nourished?

d. What theological insights can we draw from the way the fisherfolk mold their religious faith and practices in their everyday life?

The design of this study is based on three hypotheses that give it coherence and a direction. First, the fisherfolk have an organic relationship with the ocean and all marine life that enables them to see the ocean not as a place of danger but one of nourishment and life support. Second, the fisherfolk are a deeply religious people, firm in their faith convictions and devoted to their ritual practices, and this religiosity sustains them before risk and danger. Third, the Catholic fisherfolk have incorporated elements from their neithal religious heritage to mold their own Catholicism to suit their life context.

Study of lived religion calls for appropriate methods that are quite distinct from the classical methodologies of dogmatic theology and seeks the help of various disciplines, particularly the social sciences. Beliefs and practices are not codified unlike in creedal religions and so can be approached only indirectly. An ethnographic method was found more appropriate as the literacy rate among active fishermen was marginal. The study is interdisciplinary in nature and sociotheological in perspective. I have made every attempt to allow the coastal reality to speak for itself even when theoretical perspectives from different disciplines are brought into the analysis. Data gathering was done in two phases. The first phase was completed during 1995 and 1996 and relied on participant observation, in-depth interviews with informants, focus group discussions, family histories, parish records, and other historical documents. Supplementary field surveys were used to cross-check and enrich the data gathered from other sources. The findings of the first phase became the subject for the earlier version of the book published in 2001.

The same coastal context was revisited in 2018 soon after Cyclone Ockhi hit the Thiruvananthapuram coast. The impact of the disaster was visible all along the coast, and that formed the background for gathering fresh data during August through December 2018. In-depth interviews were conducted with some fishermen who had survived Ockhi, and also with the relatives and colleagues of those who lost their lives in the disaster. Focus group discussion with groups of fishermen, activists, and student volunteers helped to bring diverse perspectives together. Many changes in the coastal communities were visible, but the way people live their religion did not seem to have changed significantly. The findings of both the studies have been incorporated in the present edition of the book.

Some limitations of this study need to be acknowledged. The first is the geographical limitation as it relied on ethnographic data gathered from a restricted area of the coastal region. Though many of the findings reflect the shared experience of fishermen elsewhere, it cannot claim to speak for them. The second is the religious factor. The study confined itself to the Catholic fisherfolk who form the vast majority of the fishing population in the district of Thiruvananthapuram. Some Muslim and Hindu fishing hamlets lay scattered in the district, and data have not been gathered from these. The third is the gender factor. Even though the book speaks of the fisherfolk in general, what it contains is primarily the experience and worldview of the sea-going fishermen. I am aware that there exists an amazing world of lived religion that belongs specifically to coastal women with their own beliefs and ritual practices. This area has not received due consideration in this book. Two reasons could be given for this. Primarily, my access to this area was restricted. Besides, though women share the occupational rhythm of the menfolk to a large extent, their religious world is different as they are kept away from fishing proper and engage only in the marketing of fish and allied occupations. This area remains open for more competent scholars to explore.

The book is organized into nine chapters. Chapter 1 presents the problem to be studied starting with the crisis period soon after Cyclone Ockhi and examining the role played by religious institutions during the crisis. The chapter explains the objectives of the study and the methodology being followed. Chapter 2 takes a look at the coastal studies in the Kerala context during the past few decades and gives an overview of the fisheries sector and the fishing communities with particular reference to the Christian fisherfolk of the Thiruvananthapuram coast. Chapter 3 discusses the religious roots of the Christian fisherfolk. Their Christian history begins with Francis Xavier, the Jesuit missionary, in the middle of the sixteenth century, but prior to that they had inherited what could be termed Indic/neithal religiosity. The chapter examines how the fisherfolk negotiated the two religious traditions as they moved from neithal religion to a creedal religion. Chapter 4 looks at the neithal worldview more closely and highlights the role of the ocean and the influence of celestial bodies on their daily life and occupation. *Kaṇiyam*, the astronomical knowledge of the fishermen is discussed in detail as it is vital to fishing and to the safety of fishermen.

It is impossible to demarcate the religious sphere from the secular sphere in the case of the fisherfolk. Therefore the study deals with the coastal rituals as a window to get a glimpse of their lived religion, and this is the concern of Chapter 5. Fishing itself can be seen as a ritual performance that embodies their religious outlook. Mantric rituals are used widely to seek blessings and to ward off evil. Chapter 6 tries to connect these diverse elements to see how the fisherfolk construct their lived religion. Cintāthira Māta, the patron of the church at Vizhinjam, is a rich symbol that explains how the coastal people affirm their Catholic faith drawing nourishment from their neithal religiosity. A distinguishing mark of Christian saints who are widely venerated in coastal villages is that they all display militant features. Chapter 7 gives an overall synthesis of the lived religion of the coastal people and names it Neithal Catholicism. The two streams of Iberian Catholicism and Indic/neithal religiosity engage in a process of encounter and dialogue for this outcome, and it becomes the lived religion of the fisherfolk. They value both the streams as complementing each other; one stream gave them structural identity while the other provided a meaning frame. The chapter utilizes the Indian *rasa* theory to highlight the reconciliation of ambivalent features not only in Mother Mary but also in warrior saints.

Chapter 8 is an examination of some structural aspects of Neithal Catholicism. It notes the importance given to the confraternities and pious associations as if these are cultural substitutes in a post-conversion scenario. The chapter also examines the all-powerful role of the Vicar, the disappearance of the traditional leadership roles like Pradhāni and the crisis of leadership that the present generation has been experiencing. The general conclusion provides a summary of the major findings of the book and indicates how these could challenge religious studies and the discipline of theology. It also mentions some areas that need serious theological consideration. My hope is that this book will contribute to enriching Christian life on the seashore and promote theological reflection, taking lived religion with the attention it deserves.

CHAPTER 2

MARINE FISHERIES AND THE TRADITIONAL FISHERFOLK

Interest in Coastal Studies on the Rise

The coastal context has witnessed a growing academic interest in the traditional fishery sector since the 1960s, resulting in a large volume of research studies and publications. This was particularly significant for the Kerala coastal scene. Many of these studies were concerned either with developmental economics or with fisheries management; some were specifically technoeconomic in nature.[1] The complexity of the fisheries situation and the increasing misery of the traditional fisherfolk often led to conflicts and political agitations. This resulted in the appointment of a series of commissions by the State government to study the issues and make recommendations.[2] The major thread running through these reports and publications was the introduction of new technologies and their impact/consequences on the fisheries economy and the fishing communities in the State. The Indo-Norwegian Project that was started in 1952 can be considered the benchmark by and large. Other scholars,

1 See Jean Philippe Platteau et.al., *Technology, Credit and Indebtedness in Marine Fishing—A Case Study of Three Fishing Villages in South Kerala* (Delhi: Hindustan Publ. Corporation, 1985); K.T. Thomson, "Political Economy of Fishing: A Study of an Indigenous Social System in Tamilnadu," PhD dissertation, University of Madras, 1989; John Kurien, "Technical Assistance Projects and Socio-Economic Change: The Norwegian Intervention in Kerala's Fisheries Development Experience," *Economic and Political Weekly* 20, nos. 25&26 (January 1985), etc.

2 Babu Paul et al. *Report of the Committee to Study the Need for Conservation of Marine Fishery Resources During Certain Seasons of the Year and Allied Matters* (Thiruvananthapuram: 1982); Kalawar A.G. et al., *Report of the Expert Committee on Marine Fisheries in Kerala* (Thiruvananthapuram: 1985); N. Balakrishnan Nair et al., *Report of the Expert Committee on Marine Fishery Resources Management in Kerala* (Thiruvananthapuram: 1989); N. Balakrishnan Nair et al., *Report of the Expert Committee on Ban on Monsoon Trawling in Kerala* (Thiruvananthapuram: 1991), etc.

however, tried to study the fishing communities from the perspective of culture or social change. These were mostly micro studies of specific fishing communities. Most of these researchers emphasized the vital role religion played in every sphere in the life of the fisherfolk including the economic and the political. Some studies dealt with aspects of religious beliefs and practices of the fisherfolk more specifically. Yet, a comprehensive study of the lived religion of the fisherfolk has not been attempted as is evident from the collection of published materials.

The Indo-Norwegian Project (INP) that was started in 1952 at Saktikulangara/ Neendakara in Kerala with a focus on fisheries development provided the context for the work of Arne Martin Klausen.[3] Though the project had been technoeconomic in nature, Klausen's interest took him to a social-anthropological analysis of the villages where the project had been launched. The study is "an analysis of the reaction of a few local communities to various technological-economical stimuli." At the same time, it is also a contribution to the ethnographical knowledge of an unexplored field—i.e., the relation between the Hindu communities known as Arayās and the so-called Latin Catholics in south of Kerala. Taking into account the religious factor, he tries to find out the structural characteristics of these two communities that have brought about an economic growth. He discovers that the reaction of Latin Catholics to fishing is adequate, while it is less adequate to the new distributional system. The reaction of the Arayās is inadequate to mechanization but adequate to the contributions of INP in health sector and administration. Klausen brings out forcefully the relevance of knowing the "value-transformation system" of a community in any project aimed at social change and shows how its inadequacy has hindered the INP from achieving its goals. The same insight is relevant to any change-oriented strategy—whether of a social activist or a religious agent.

Closely related to the Indo-Norwegian Project is the work by Leela Gulati.[4] It focuses on the economic and demographic impact brought about by the major changes—technological, economic, and demographic—introduced by the Project over three decades. She takes up three fishing villages for a case study, with a specific focus on women.

---

3 See Arne Martin Klausen, *Kerala Fishermen and the Indo-Norwegian Pilot Project* (London: Allen and Unwin, 1968). The Indo-Norwegian Pilot Project was started in 1952 in the twin fishing villages of Saktikulangara and Neendakara in the Kollam district of Kerala aimed at fisheries development through modernization of technology. This project introduced fully mechanized trawler boats in the State.

4 Leela Gulati, *Fisherwomen on the Kerala Coast* (Geneva: International Labour Office, 1984).

The detailed case histories of ten women help us to get an insight into the inner world of fisherwomen. The book highlights the importance of taking the women's dimension fully into account in any developmental program.

P.R.G. Mathur has given us one of the pioneering yet comprehensive ethnographic records of any fishermen community in Kerala.[5] It is a study of the Mappila fishermen of Tanur in Malappuram district and is an attempt to understand the structural relations between the habitat, technology, economy, society, and culture. The Mappila fisherfolk of Tanur are native converts to Islam. Mathur analyzes a number of rituals and magico-religious practices prevalent among them and argues that these fishermen, though Muslims by religion, still retain many cultural and religious elements from their pre-Islamic tradition, namely that of Hindu fishermen. They even retain their coastal Hindu clan names. Traditional ideas about certain gods and goddesses and beliefs in the spirit of the departed are retained. Weaving all these together, Mathur concludes: "Islam has been superimposed upon a primitive culture... Thus, we find that there are coexistence and integration of the beliefs and practices of the Hindus with the Islam tradition. The local tradition is very much syncretised with the universalism of Islam."[6] The account of Mathur is significant because of two reasons. First, it focuses on a fishing community that had changed their religion in the past, very similar to the Catholic fishing community in Trivandrum. Second, the preconversion roots are also very similar in spite of the geographical distance. Mathur has indicated a fruitful way to study the complex world of coastal religion. Some caution is required since he tends to reduce whatever is non-Islamic in their ritual life as *Hindu*. Other cultural traits like Jainism, Buddhism, as well as Pre-Dravidian tribalism have also richly contributed to the present-day coastal culture.

Two studies concentrate specifically on the fisherfolk communities around Vizhinjam.[7] Velacherry has analyzed the community from 1900 till 1982 with a focus on the social conflict by applying the Weberian model of stratification at the micro level. He took up the two main fishing groups at Vizhinjam, viz., the Christians and the Muslims, and narrated a total of 37 cases of conflict in the village, 31 of which were

---

5 P.R.G. Mathur, *The Mappila Fisherfolk of Kerala: A Study in Inter-relationship between Habitat, Technology, Economy, Society and Culture* (Thiruvananthapuram: Kerala Historical Society, 1977).

6 Mathur, *Mappila Fisherfolk,* 350.

7 Joseph Velacherry, *A Study of a Fishing Community in Kerala* (unpublished PhD Dissertation, University of Delhi, 1982). T.K. Vimalakumari, *Infant Mortality Among Fishermen* (New Delhi: Discovery Publishing House, 1991).

intercommunal. He traced how the former relationship of domination and subjugation gave way to relationships governed by principles of equality and coexistence. Vimalakumari has inquired into the causal factors of infant mortality at Vizhinjam comparing the same religious communities. But she does not explore the religious factor any further. The work by Patrick A. Roche is helpful in comparing the Mukkuvar of the Thiruvananthapuram coast with the Paravas of the Coromandel Coast in Tamil Nadu.[8] The Paravas were converted to Catholicism by the Portuguese in the sixteenth century, some years before the conversion of the Mukkuvar, and the book is a study of their transformation during the period CE 1860–1960. Roche explored how Roman Catholicism emerged as a base of identity and affirmation among the Paravas during the Portuguese period. He explained how Catholicism gradually replaced the Hindu idols, Brahmin rituals, and temple festivals by introducing statues, rituals, and festivals of a Christian color, thus creating a Parava Catholicism that is partly Iberian and partly Tamil in flavor.

One study that took the whole of the Kerala coast as its canvas with focus on the religious factor is by Houtart and Nayak.[9] Acknowledging the central role religion played in strengthening the differentiation among fishermen, they studied the three dominant religious communities, Hindus, Christians and Muslims, in three districts of Kerala, surveying twenty villages and interviewing 750 persons. It highlighted the Christian community's extreme dependence on the religious institution and the central place of the church in their entire social life. It concluded saying, "Being a minority, who was marginalized and vulnerable...it had to affirm its identity that was best defended by the religious institution... . Exterior signs of identity were important, and the function of the totem was replaced by the impressive church buildings erected all along the coast."[10] The book is certainly a bold attempt at understanding the "cultural universe" of the fisherfolk and the central place of religion in it. However, the comparative approach of the book yields only negligible result since religion is taken out of its natural setting and the worldview of a community. The study recognized the impact of the pre-capitalist forms of existence on the shaping of ideas and representations of the fisherfolk, but without sufficiently relating it to their tribal background before the nineteenth century. Probably this is a limitation of any purely sociological method if applied to a complex realm like culture and religion.

---

8 Patrick A.Roche, *Fishermen of the Coromandel: The Social Study of the Paravas of the Coromandel* (New Delhi: Manohar, 1984).

9 Francis Houtart and Nalini Nayak, *Kerala Fishermen: Culture and Social Organization* (Thiruvananthapuram: PCO Centre, 1988).

10 Hourart and Nayak, *Kerala Fishermen,* 71.

A monograph by Nalini Nayak is a complementary volume to Houtart and Nayak (1988) and is a study on the role and participation of women.[11] It is a collection of case studies of five fishing villages along the south-western coast, and four of them are in Kerala. It tried to investigate the socioeconomic aspects of the artisanal sector following the stage of motorization in the 1980s. While assessing the impact of social institutions on women's consciousness, the study underscored the central role of the religious institutions in their lives. The study suggested that "a deeper understanding of the role of religious institutions, practices, and taboos in determining the worldview of the fish workers will help understand further how the community sustains itself."

Kalpana Ram has explored the perspective of coastal women in greater detail.[12] The book is about Mukkuvar women with reference to social change and capitalist transformation. It has well integrated both the theoretical concerns of Indian sociology and the ethnographic rigor of an empirical study. The focus is on the cultural construction of meaning touching upon caste, the landscape, the female body, as well as the labor process. Ram studied the lives of Mukkuvar women as social subjects in an occupationally polluted community and showed that an investigation of the religious framework is essential in order to evaluate Mukkuva ideas of the female body. Discussing the religious consciousness of the Mukkuvar, she threw light on the tensions that existed not only between them and the Hindus but also between popular Catholicism and the official church. She illustrates, referring to the cult of the Virgin Mary, the way popular Catholicism of the Mukkuvar has borrowed from certain strands of Hinduism overstepping the boundaries of the institutional church. Ram, however, did not venture into the other aspects of their religious consciousness.

A study that dealt with the Hindu fishing community of Dhīvara is that of Udayabhanu.[13] The term Dhīvara has come into current use only recently to refer to the conglomeration of all Hindu castes engaged in the fishing occupation.[14] It was in 1984 that the government accepted the

---

11 Nalini Nayak, *Continuity and Change in Artisanal Fishing Communities: A Study Report* (Thiruvananthapuram: P.C.O. and SIFFS, 1993).

12 Kalpana Ram, *Mukkuvar Women: Gender, Hegemony and Capitalist Transformation in a South Indian Fishing Community* (New Delhi: Kali for Women, 1991).

13 K.M. Udayabhanu, *The Dhīvara Community and Social Change in Kerala* (unpublished PhD Dissertation, University of Kerala, 1990).

14 In fact, Dhīvara is an ancient term coming from the Vedic period. Dhīvara and Kaivarta were fishermen groups probably counted among the *dasa* of the Ṛg Veda.

term officially to include all these subcastes in the OEC (Other Eligible Communities) category. The Dhīvaras of Kerala consist of Arayās, Mukkuvas, Nuḷayas, Mugeyas, Paravas, and others. Udayabhanu places the Dhīvara community within the broad matrix of social change that has been taking place in Kerala for over a century. This work does not substantially contribute to an understanding of the culture or religion of the fishing communities.

Some other works are specifically technoeconomic in nature and so are not taken up for discussion.[15] More recent studies tend to focus on specific aspects of coastal life, culture, or occupation. The unpublished thesis of Benny Chiramel is a recent study from a psychosocial perspective.[16] More and more studies and reports made by the new generation of scholars from within the fishing community is a welcome development. This stands in contrast to the earlier contributions that were mostly by scholars from outside the community. The reason for this promising trend could be the significant progress in general education that the community has achieved in the past few decades and the growing number of scholars with academic credentials. Robert Panipilla has succeeded to incorporate the unique coastal vocabulary into his book that is a fine blend of both coastal experience and scientific inquisitiveness.[17] Some coastal organizations and interest groups seem to make conscious efforts to promote coastal studies and related pursuits.[18]

In short, we can say with confidence that a considerable amount of work has already been done on various aspects related to the seashore and the coastal people during the past few decades. The major conclusions arrived at by the researchers can schematically be summarized as follows:

- The religious factor plays a predominant role in the identity formation of the fisherfolk. It is difficult to understand their cultural universe except in relation to religion which plays a major influencing role in their lives.

---

15 Examples are Jean Philippe Platteau et.al., *Technology, Credit and Indebtedness*; K.T. Thomson, *Political Economy of Fishing.*

16 Benny Chiramel, *The Phenomenon of Social Exclusion of Fisherfolk Youth in relation to their Self-esteem, Aggression and Religious Identity* (Unpublished PhD Dissertation, University of Mumbai, 2014).

17 Robert Panipilla, *Kadalarivukalum Neranubhavangalum* (Malayalam) (Kottayam: Current Books, 2015).

18 Friends of Marine Life, Coastal Higher Education Society, Coastal Watch, Coastal Students' Cultural Forum, etc. are among the new groups/associations that have emerged on the Thiruvananthapuram coast.

- The religious conception of the fisherfolk seems to be constructed basically on their relationship to the sea rather than to the land; in other words, the marine habitat and the fishing occupation contribute significantly in shaping their religious worldview.
- Even though the fisherfolk along the Kerala coast have been converted to Christianity or Islam during the past few centuries, religious elements of the preconversion heritage persist in their present-day religion as they live it.

The present book bases itself firmly on these conclusions. However, these studies do not seem to succeed in perceiving the inner world and thought patterns of the fisherfolk, and the influence of these on the process of organizing and orienting their daily life. This has happened because these studies have not paid enough attention to the lived religion of the coastal people. The academic gap in this area is a fact and is felt deeply by planners, researchers, as well as change agents. The present inquiry is a partial response to this urgent need and intends to build on the findings of the past.

Marine Fisheries—The Broader Context

India is the second largest fish producing nation in the world, with a share of 5.4 percent of global fish production. It provides livelihood to nearly 0.874 million fishermen families in the country. With a coastline of 6068 km spread across 9 States and 3288 fishing villages, India had a fisher population of 3,999,214 in 2010.[19] The fleet of fishing craft consists of 37 percent mechanized boats that use machine power for propulsion and fishing operation, 37 percent motorized boats or catamarans, i.e., fitted with an outboard engine only for propulsion, and 26 percent nonmotorized crafts. These are small-scale fishers who catch fish in near-shore waters using labor-intensive technologies. The depletion of fishery resources during the past few decades has brought in a new awareness that fish is no more an inexhaustible resource, and that regulations are necessary for effective management of the marine fisheries. While many countries have moved ahead in developing aquaculture commercially, India has been lagging behind. The States with the largest fisher population are Tamil Nadu, Orissa, and Kerala. In all the States except Kerala, the dominant religious group is Hindu (76 percent) followed by Christians (15.2 percent) and Muslims (9 percent). In the case of Kerala, the Christian fisherfolk is the largest with 42.7 percent followed by Hindus (29 percent) and Muslims (28.3 percent). It is worth noting that 92 percent of

---

19 Ministry of Agriculture, Government of India, and CMFRI, *Marine Fisheries Census 2010, India, Part I* (Kochi: CMFRI, 2010), 27.

the Christian fisherfolk live and work in the two southern states of Kerala and Tamil Nadu.[20]

Among the Indian States, Kerala is the second largest contributor to marine fish production ranked after Gujarat and contributes about 23 percent of the all-India marine fish production. Forty percent of the sea-food export earnings are credited to Kerala. The population of the State according to the Census of 2011 is 33.4 million, and 3.1 percent of this population is estimated to be engaged in fishing or allied occupations. The State Planning Board points out that a higher percentage of the fish-erfolk in Kerala is in the marine sector unlike in the rest of India. If the ratio of marine and inland fishermen in India is 35:65, the order is re-versed in the case of Kerala, where the ratio is 71:29.[21] Export of marine products is a major source of income. Frozen shrimps form the most at-tractive item for export.

Kerala has a coastline that is 590 km long, and two vigorous mon-soons accompanied by the unique phenomenon of mud banks (*cākara*) along the coast. Upto 222 fishing villages and 12 fishing harbors dot the coastline of Kerala. Even though Kerala's share in the total population of India is only 3.7 percent, more than 30 percent of India's marine fish-ermen live in this State. According to a census in 2010, the State had 1,18,937 fisherfolk families and a fisher population of 6,10,165. Of these, 1,18,937 were active fishermen, full-time or part-time, operating 12,781 fishing crafts. Upto11,175 of these crafts are motorized boats or catama-rans (fitted with outboard engines) while 4,722 are mechanized vessels, and the rest are nonmotorized canoes or catamarans.[22] Catamaran is a raft made of four or five logs of softwood, usually of Albizzia Stipulata, yoked together with coir ropes, and used mostly in the southern districts. The recent trend is to replace these with boats made of plywood or fi-breglass and fitting them with outboard engines. The mechanized sector (trawling boats, gillnetters, purseiners, etc.) absorb the rest of the fishers and is concentrated at major ports like Kochi and Neendakara. A majori-ty of fisherfolk families owned neither fishing crafts nor gear. Traditional fishermen have been categorized under unorganized labor, and even *unorganizable* by trade unions mainly due to the erratic nature of the occupation. As compared to other working-class segments in the State, unionization of artisanal fishermen has only a short history. However,

---

20 *Fisheries Census 2010, India, Part I*, 33.

21 State Planning Board, *Economic Review 2016–Agriculture and Allied Sectors, Fisheries Development* (Thiruvananthapuram: State Planning Board, 2016).

22 Ministry of Agriculture, Government of India, *Marine Fisheries Census 2010*, Part II, 6, Kerala (Kochi: CMFRI), 1–3.

issues like bottom trawling, deepsea fishing, etc. have contributed to their speedy mobilization in recent times.[23]

The Fishing Community in Kerala at a Glance

Traditionally, the fisherpeople belonged to the lowest segment of the society, usually outside the caste hierarchy, and the very occupation was considered polluting. A large number of them got converted to Islam or to Christianity, although the occupational stigma persisted. The Arab traders introduced Islam on the Malabar Coast in the eighth century. The Mappila Muslim fisherfolk of north Kerala form a cultural community with its own unique features. The history of Christian fishing communities begins with the arrival of the Portuguese at the turn of the fifteenth century; this aspect will be taken up for discussion later.

The religious division of Kerala's fisherpeople has been noted for the comparatively high proportion of Christians and Muslims to Hindus. The population pattern at the all-India level shows 83 percent Hindus, 10 percent Muslims and 2.6 percent Christians; but the picture of Kerala stands in contrast with 54.73 percent Hindus, 26.56 percent Muslims and 18.38 percent Christians (2011 census figures). A more balanced proportion is present among the fishing population of the State with 29 percent Hindus, 28 percent Muslims, and 43 percent Christians.[24] These religious groups are further divided into endogamous subcastes with little mutual interaction. The Hindus include subcastes like *Arayan, Vālan, Marakkān*, and *Nuḷayan*. They have been given the generic name Dhīvara in modern times. Dhīvara is an ancient term coming from the Vedic period. Dhīvara and Kaivarta were fishermen groups, probably counted among the *dāsa* of the Ṛg Veda. The term has come into current use only recently to refer to the conglomeration of all Hindu castes engaged in the fishing occupation. It was in 1984 that the government accepted the term officially to include all these subcastes in the OEC (Other Eligible Communities) category. In the northern districts of Kasargod, Kannur, Malappuram, and Trichur, there are high concentrations of Muslim fisherpeople, while in the southern districts of Kollam and Thiruvananthapuram, Christians form the major segment; both Hindu and Christian fishers reside in Alappuzha and Ernakulam districts. The Christian fisherfolk in Kerala generally belong to the Latin Catholic denomination.

---

23 See *Oru Samara Katha* (Thiruvananthapuram: Kerala Swathanthra Matsya Thozhilali Federation, 1988). It narrates the story of the State-wide fishermen's agitation in 1984-85 demanding new legislation for protection of fishery resources and welfare of the traditional fishermen.

24 *Fisheries Census 2010*, Part II, Kerala, 3.

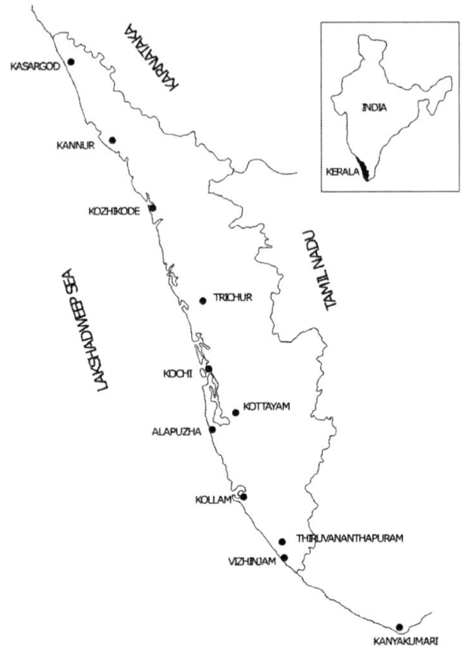

Fig. 1. Kerala State—Location Map showing Thiruvananthapuram coast

The district of Thiruvananthapuram with 78 km of coastline has the largest fishing population in the State residing in 42 fishing villages. The 2010 census figures give the total number of the fisherfolk as 1,46,326 of which 30,879 are active full-time fishermen.[25] The religious composition is also noteworthy. Christians formed the vast majority with 27,944 persons (83.8 percent), the Muslims coming next with 4788 (14.4 percent) and the Hindus forming a small minority with 608 (1.8 percent).[26] Panathura and Kovalam are Hindu fishing hamlets, while Perumathura, Beemapally, and Vizhinjam North are Muslim pockets. Some traditional fishing villages of Hindu fishermen (Arayās) are no more considered fishing villages since people do not engage in the traditional occupation of fishing any more. Fisheries in this district take the credit for many special features. The sea is rough, and so the use of catamarans is widespread as it is best suited for rough seas. Mechanized crafts are totally absent and 2880 fishing crafts are motorized while 2304 are nonmotorized. Thiruvananthapuram fishermen are known for their fishing skills

25 *Fisheries Census 2010,* Part II, Kerala, 9–14.

26 *Fisheries Census 2010,* Part II, Kerala, 15.

and remarkable courage that lets them go with hook and line to the deep waters in search of rare fish. This explains why a vast majority of hook and line fishermen come from this district.[27] The deep impact of these operations on their attitude and belief system is undeniable. The active participation of the womenfolk in allied activities, particularly in the marketing of fish, is unique to this district. In 2010, there were 10,362 women engaged in the marketing of fish.

Vizhinjam—A Nerve Center of Fishing Activity

The coastal village of Vizhinjam is also a major fishing harbor and is situated 18 km south of Thiruvananthapuram city at longitude 76° 59′ East and latitude 8° 22′ North. Kovalam, the well-known beach resort, is only 2 km north of Vizhinjam. The beach is quite small and narrow, but the two natural rock formations jutting out into the sea create a natural bay for country crafts as well as mechanized boats. This makes Vizhinjam a safe haven for fishing operations all through the year, especially during the rough monsoon season. Vizhinjam Sea has a natural depth of more than 20 meters within 1/2 a nautical mile. At present, the fishing harbor was traditionally being used mainly by the artisanal sector with hundreds of motorized plywood boats and plank-built canoes anchoring here during the fishing season.

Image 1. Vizhinjam fishing harbor in Thiruvananthapuram

The natural setting makes Vizhinjam a twin fishing village, and government records refer to Vizhinjam North and Vizhinjam South as two separate fishing villages among the 42 in the district. Vizhinjam South is

[27] Thiruvananthapuram district alone had 7510 hook and line fishers in 2005. See *Marine Fisheries Survey 2005*, 41.

more vibrant in fishing activity with a fisher population of 11,606 in 3060 fisher families, while Vizhinjam North has 5849 fishermen in 1050 fisher families. But the difference is more than geographical. Fisherpeople belonging to two different religious communities populate either side of the bay. Muslims inhabit the northern side, called Mathilpuram, while the southern coast, known as Kottapuram is inhabited by the Latin Catholic fisherfolk. According to the 2010 fisheries census, all fisher families in Vizhinjam belonged to the Muslim fold with no Hindu or Christian fishermen residing there. Vizhinjam South, on the other hand, had 3054 Christian families with just three Hindu families and three Muslim families.[28] The road leading from the fish market to the seashore categorically separates the two communities, both in terms of housing as well as fishing operations. Both communities have their places of worship prominently located within their respective areas. The Catholic fishermen on the southern side are known to be dynamic and daring, operating even beyond 100 meters of depth, using a variety of gears, while the Muslim fishermen on the Northern side confine themselves to shallow water operations. Vizhinjam South witnesses intense fishing activity all through the year. The amazing courage and valor of the fishermen here enable them to undertake risk-ridden fishing operations even in rough weather.

Vizhinjam was well known as a place of commercial and military importance from the very early times. The literature of the Sangam period refers to three dynasties ruling the geographical area known as Kerala today—the *Āy* dynasty ruling the southern part, *Nanna* dynasty in the north, and *Chēra* dynasty in the middle, indicating a gradual transition from a tribal setup to kingly rule. The *Āy* dynasty ruled the southern tip of India up to the tenth century BCE and Vizhinjam was fortified as the capital of many *Āy* kings.[29] Chatayan, the *Āy* king, is believed to have his capital at Vizhinjam.[30] The historic Paliyam document is considered as written by VikŗamādityaVaraguṇan, the *Āy* king ruling from Vizhinjam; it refers to his offering a large amount of land to the Buddhist institution of Srimūlavāsam.[31] After VikŗamādityaVaraguṇan, the *Āy* dynasty declined and Vizhinjam was captured and later destroyed by the *Pāndyas*. The period after the ninth century saw Vizhinjam becoming the venue of many battles between the kings in southern India—the *Chēras*, the *Pāndyas*, and the *Chōlas*. In the twelfth century, Vizhinjam was renamed

28 *Fisheries Census 2010,* Part II, Kerala, 135

29 Government of Kerala, *Kerala State Gazetteer*, Vol. II, Part I (Thiruvananthapuram: 1986), 155.

30 Desabandhu Publications, *Kerala Vijñāna Kosam* (Thiruvananthapuram: Desabandhu Publications, 1988), 58.

31 *Kerala Vijñāna Kosam*, 59.

Rajendra Chōḷa Paṭṭaṇam by the *Chōla* king Rajendra Chōla Deva, ruling from Kancheepuram. This name is still seen in old deeds and documents.[32] The renowned goldsmith community (*Thaṭṭān* or *Viśwakaṛma*) in Vizhinjam, occupying the north street (*Thattatheru*) claims to have been brought here from Kancheepuram centuries ago by the then rulers of Vizhinjam.

Place names at Vizhinjam clearly indicate the existence of a royal past. *Mekkōṭṭa* (upper fort) is believed to be the location of the ancient fort on a hillock; locations like *ānakkuzhy* (elephants pit) and *kuthirappanthy* (horses' line) point to the military past of Vizhinjam. *Kariyilakkeṭṭu* comes from *karu-ōlakkeṭṭu* (storehouse of gunpowder), and *ūṭṭukālparambu* is said to be the place where meals for the kings were prepared. *Gaṇapathy Viḷākom* was the venue of the temple of Gaṇapathy, the reigning deity of the kingdom. According to a local tradition, there was a stone inscription at the temple site, referring to its owning 100 acres of field at Kadiyapattam. *Thoppāsikudy* is the area where *thoppāsis* were settled; *thoppāsi*, in Tamil, refers to those who speak two languages, say Malayalam and Portuguese, and so could act as interpreters (*dvibhāṣis*), very much in demand in a port city carrying on trade with the Arabs, and later with the Portuguese and the English. *Kastūrikuḷam* is said to be the pond for the bathing of royal women. The royal palace is believed to have existed at *Chenkāli* (*chenkōl* means scepter); a major part of this site is under the sea today. The palace complex is said to be lying under the present beach beneath 34 feet of sand. People used to collect gold granules and pieces, presumed to be parts of the ornaments of images, from the beach sand until recently, thus strengthening the possibility of a submerged royal past. Recent archaeological excavations seem to indicate that Vizhinjam probably had the earliest fort-remains that exist in Kerala today. After noting that Vizhinjam was a fortified metropolis during the medieval period, Ajit Kumar writes: "Apart from being a political stronghold, pottery finds indicate that it was also an active and strategically located international port of call in the Indian Ocean trade with catchment areas of the trade extending (from) the Mediterranean or the Middle East to China in the Far East."[33]

Vizhinjam did not lose its importance even after the arrival of the Portuguese. Miguel Ferreira, the Portuguese captain of the Coromandel Coast, burned down the village of Vizhinjam on his way to Ceylon, since

---

32 Nagam Aiya, *The Travancore State Manual*, Vol. I (New Delhi: Asian Educational Services, 1989), 251; *Kerala State Gazetteer*, Vol. II, part I, 186.

33 See Ajit Kumar, "An Appraisal of Vilinjam's History and Archeology in the Wake of Recent Explorations", *Kerala Spectrum—Aspects of Cultural History*, ed. Charles Dias (Cochin: The Indo-Portuguese Cultural Institute, 2006), 79.

the Muslim inhabitants of Vizhinjam prevented him from landing and getting water.[34] This happened in 1539. Records show that Mukkuvar had been converted to Islam earlier under Arab influence and that there were Muslim fishermen at Colachel, Thengapattanam (Taingapatam in missionary documents) and Vizhinjam when Francis Xavier reached these villages. Some of the coastal people converted by Xavier were probably former Muslims.

Nagam Aiya writes about Vizhinjam:

> This was the first English settlement in Travancore. It was one of the earliest centres of trade and has played an important part in the early history of Travancore. During the tenth and twelfth centuries, the town and the country round about came under the sway of the *Cholas*. It was subsequently a Dutch factory, and in 1505 it was sacked and burned by the Portuguese. In 1644 it was granted to the East India Company for the erection of a factory. After the establishment of the Anjengo factory by the East India Company, it was relinquished by the Company, and its importance declined.[35]

We notice a chronological discrepancy about the date of burning of Vizhinjam by the Portuguese. The Dutch set up a factory and a pepper warehouse at Vizhinjam many years after the Portuguese reached there, and therefore the burning of Vizhinjam could have occurred only in 1539.[36]

Kesava Pillay who was Dewan (Chief Minister) in the princely State of Travancore from 1789 onward had realized the commercial potential of Vizhinjam as a seaport. As part of his grand plan to improve all the seaports in Travancore, he visited Vizhinjam and made arrangements for establishing a port and a large warehouse.[37] But he couldn't realize his dream. Till recently, Vizhinjam remained mainly a fishing village and a fish landing center rather than a port town or an administrative center. But drastic changes began to take place since 2015. One key factor was a new plan to develop Vizhinjam as an international commercial port, cashing in on its closeness to the international shipping route. It was actively pursued by the State government, and an agreement with Adani

---

[34] George Schurhammer, *Francis Xavier—His Life, His Times*, Vol. II, India, 1541-1545, trans. M. Joseph Costelle (Rome: The Jesuit Historical Institute, 1977), 263.

[35] V. Nagam Aiya, *The Travancore State Manual*, Vol. III. (New Delhi: Asian Educational Services, 1989 edition), 602.

[36] *The Travancore State Manual*, Vol. I, 287; see also Schurhammer, *Francis Xavier*, 559.

[37] Benjamin S. Ward and Peter E. Conner, *Memoir of the Survey of the Travancore and Cochin States*, Vol. I., rev. ed. (Thiruvananthapuram: Government of Kerala, 1994), 184.

Image 2. Vibrant life on the seashore

Ports and SEZ Private Ltd was signed for implementing the Vizhinjam International Deep-water Multipurpose Seaport project. Vizhinjam International Seaport Limited (VISL) is the implementing agency. The work has been progressing and is likely to displace many neighboring fishing villages, thus pushing the traditional fishing communities to look for other means of livelihood. However, it is too early to assess the full impact of the project on the survival of fishing communities along the Thiruvananthapuram coast.

The Catholic Fisherfolk at Vizhinjam

Since Vizhinjam was traditionally a vibrant fishing center, the population consisted of active fishermen families and people connected with fish trade or processing. The latter group includes small traders in shrimp and cuttlefish, headload women vendors, auctioneers, those dealing with ice, kerosene, and oil, and those doing the repair of boats and outboard engines. Besides these occupational groups, categorization takes place based on where people come from: *kizhakkathukār* (the easterners) refers to fishermen coming from Tamil Nadu coast, *kīḷakkār* (those from below) refers to those from the coastal areas south of Vizhinjam, and *mekkathukār* (those on the upper side) refers to people along the coast toward Kollam. *Kuṭṭakār* or *kachavadakkār* refers to all those who come from the interior villages/towns as traders or fish vendors. We may note the importance of direction implied in these usages, and we will discuss this aspect later. In terms of religious composition, we find three major religious groups residing in Vizhinjam. The term Mukkuvar is used exclusively for the Catholic fisherfolk who form the dominant group along the coast. *Chōnavan* is the term Mukkuvar used to refer to the Muslim fisherfolk; this includes both *Marakkāns* (fishermen) and *Mēthans* (traders). Today we do not find many in the Hindu population engaged in fishing. The term *Maṛujāti* is commonly used by the Mukkuvar to refer to the Hindu population irrespective of their caste. The focus of the present study is the Catholic fisherfolk.

The history of the Catholic community at Vizhinjam since the period of the Portuguese missionaries remains largely unknown. The church establishment has failed in maintaining the essential historical documents. The suppression of the Society of Jesus (to which Francis Xavier and his companions belonged) by the middle of the eighteenth century might have affected the Catholic fisherfolk deeply. The Trivandrum Mission comprising coastal villages from Pallithura to Thengapattanam continued under the Portuguese protection (*Padroado*) of the diocese of Kochi, and that link continued until as late as 1955. Increasingly, the Goan clergy had been serving the Trivandrum Mission, and the Goan hegemony too was at times resented by the coastal Catholics.[38] The old church existing on the beach has a stone inscription in Latin on its front door post, which reads as follows:

> BEATAE MARIE VIRGINI DICATA
> SUB PATRONATU LUSITANIAE REGUM
> .... ECLESIA RECTA ATQVE
> DIE 8 OCTOBRIS BENEDICTA FUIT
> REV. VIC. J.R. ATHANASIO REBELLO
> ORIUNDO EX MARGAM DE GOA 1875.
>
> [Dedicated to the Blessed Virgin Mary
> under the patronage of the Kings of Portugal
> the church was erected
> and on October 8 was blessed
> with Rev. J.R. Athanasio Rebello
> coming from Margam in Goa, as Vicar, in 1875.]

The church is still used by the local Catholics, although a new church was built at Kottapuram area recently that now functions as the parish church of the Catholic community. The Patron of the church is *Sindu Yātra Māta* (Our Lady of Sea Voyage), popularly known as Cintāthira Māta. How and when the ancient church of the Assumption became the Church of Sea Voyage is shrouded in mystery. In 1995, the parish had a membership of 9340 Catholics belonging to 2176 families.[39] This number

38 The title of a two-piece article that appeared in the *Latin Christian* paper (April 11 and 12, 1935) was "Russian Rule in the Portuguese Missions of South Travancore"; it referred to the hardships inflicted upon the coastal Christians by the Goan clergy. It also alleged that Mgr. Lino D'Souza, the then Vicar of Vizhinjam, had taken Rs. 2000/- as bribe from the Pradhānis of Thuthur for permission to reopen the church that had remained closed.

39 Supplementary survey conducted in November 1995.

has been going down since the new Port Project began to displace many fisherfolk families.

The Catholic fishing community at Vizhinjam and the adjoining coastal villages may appear to a visitor as a homogenous social entity consisting of seagoing fishermen and their families, all professing the Catholic faith. In fact, this picture is quite deceptive. The internal structure and dynamics of the community are complex and difficult to grasp. A careful look would reveal the existence of three major subgroups: a) *kudumbakkār* (elite families with lineage) b) *pulikiyar* (dependent workers), and c) *nāvier* (the barber group). Each group can be distinguished by the characteristic features it displays. Traditionally, the *kudumbakkār* were the owners of fishing craft and gear, especially *kampavala* (shore-seine), the major gear used in those days, the investment for which was too heavy for an ordinary family. The money power was usually associated also with social status and political power, symbolized in surnames like Gomez, Fernandes, Morais, Gonsita, Gera, Roche, and D'Costa. These surnames have come from the Portuguese patrons who became godfathers to the converts at baptism, explains one informant. The Pradhānis (community leaders) came from these families.

The history of the past half a century is very much centered on the complex relationship between half a dozen elite families with shifting alliances and conflicts. The major families are Mantri kudumbam, Mankaḷāvuveed, Kuñchuveed, Thakkāṭṭuveed, and Kōlāchiveed. In course of time, there arose new elite families like Vḷāvathu, Vēḷāḷachāru, and Veḷḷithuṭṭu. There is also a status distinction between those who are natives and those who have settled from elsewhere (*varathar*); Thakkāṭṭu, Vēḷāḷachāru, etc., belong to the latter group. Occasionally we may hear someone in a fight shouting contemptuously, "What do you know? You are only a *varathan*!" Around the middle of this century, the ruling family was Mantri kudumbam. There are many stories about how they acquired this name. This family ruled till 1960 when they were overthrown by Kuñchuveed; they ruled for about fifteen years, and around 1975, Thakkāṭṭukudumbam became dominant and ruled in alliance with Thamayanveed. By late 1980, Kōlāchyveed and Vēḷāḷachāru*veed* gained prominence through family feuds. The intracommunity clash of 1993 is a clear illustration of the lineage-based power struggle within the community.[40]

40 The clash of 1993, in which one person was killed, took place between two groups of families within the Mukkuva community at Vizhinjam. The incident was centered on the house of a widow whose right of occupation was disputed. The controversy brought into the picture some leading families of the village

The *pulikiya* group consisted of people who worked as laborers in the boats owned by the *kudumbakkār*. They usually had no assets of their own. They were called *vaḷarthumakkaḷ* (foster children) by the *kudumbakkār*, thus expressing the dependency relationship. Most of them had come from Tamil Nadu and settled at Vizhinjam for work under the *udampadi* or contract system. Though the term *pulikiya* is not in common use today, it can be heard occasionally being used to refer to the worker group, often as a term of insult. This group is numerically the majority, and the boundary line is often not rigidly marked. The emerging culture of love marriages and the growing number of boat owners (made possible by the plywood boat/outboard engine revolution) could be seen as one main cause for this change.

Nāviers are not strictly fishermen, yet an essential segment of the coastal Catholic community as a dependent group assigned with the polluting activities in the community. They are variously called *nāviekkār, ambiṭṭan*, or barbers. They work primarily as barbers, but also offer various ritual services connected with marriage and death. For these services, they get a share of the daily catch from their patrons in cash or in kind. Their presence on the seashore is noticeable as they move around with a *kachān* (collection kit) to get their customary share of fish (*nāviappadi*). Velacherry records that there were 127 *nāviers* belonging to 24 families (13 households) in 1976.[41] In 1995, the number increased to 155 members belonging to 40 families (16 households).[42] A patron–client relationship of dependency is evident, and the system of a *jajmāni* relationship was noted earlier by other researchers.[43] *Nāvichi* is the term used to refer to *nāvier* women. Their occupational identity is affirmed by a saying among women, "*Nāvichi mudi veṭṭiyāl nāṇamkeṭṭu vaḷarum*" (if a girl's hair is cut by *nāvichi*, it will grow shamelessly!) The Hindu Arayar fishermen use the term *kāvuthiyan* to refer to the barber caste among them; prescribed priestly and purifying functions have been assigned to them. Among the Muslim fishermen, they are called *osāvu*. There is a reference to *nāvier (nāvitār)* in the ancient Sangam literature as people who were doing barbers' work. This is proof enough to realize their antiquity. The custom of endogamy and the low status of their work

joining either side. In the clash that followed, Stephen, a young man from the Thamayan family got killed on February 8, 1993. Fearing retaliation, all members of the Thakkāṭṭu family quit Vizhinjam and took temporary refuge in neighboring fishing villages. A police camp was set up to prevent further clashes.

41 Velacherry, *Study of a Fishing Community*, 67.

42 Supplementary survey was conducted in 1995.

43 See Velacherry, *Study of a Fishing Community*, 70; Kalpana Ram, *Mukkuvar Women*, 79.

might have kept them as a dependent subcaste within the fisherfolk for centuries.[44] Changes are visible, mainly due to the spread of education and other extraneous factors.

---

44 In the district of Kanyakumari in Tamil Nadu, *nāviers* are called *Kerala Mudaliār*, and remain a service caste for other castes like Mukkuvar and Nādār.

CHAPTER 3

THE RELIGIOUS ROOTS OF THE FISHER PEOPLE

The religious background of the Catholic fishing community along the Thiruvananthapuram coast is the subject matter of this chapter. In the first part, we discuss the Indic roots of their religiosity that is the common heritage they share with all the fisherfolk along the Kerala coast irrespective of their religious affiliation. The second part would be a more focussed inquiry into the religious past that is specific to the Catholic fisherfolk. Such an inquiry is necessary to grasp the religiocultural worldview of the coastal people that is the main concern of this study.

The Indic Roots of the Marine Fisherfolk

History shows that the fisherfolk of Travancore coast became Christians in the middle of the sixteenth century. Presently, they all owe allegiance to the Roman Catholic faith under the jurisdiction of Archdiocese of Trivandrum. Their mass conversion goes back to 1544 and was at the initiative of the Portuguese missionaries supported by the Portuguese royal power. This period is rather well documented, thanks to missionary letters and reports as well as other colonial documents. Their pre-Christian history is scanty yet sufficient to prove the existence of a unique culture and religious expression in tune with their unique geosocial environment and occupation.

The nomenclature is significant. The term *Mukkuvar*, as is used along Thiruvananthapuram coast, refers always and exclusively to people of the Catholic fishing community. The term *Marakkān* is used for Muslim fishermen, while *Nuḷayan* (also *Arayan*) is the term for Hindu fishermen. But who indeed are the Mukkuvar? How are they related to the other fishing groups? What are the special features of this group? Some preliminary clarification is necessary before proceeding further. The popular usage of the term in Malayalam does not bring more clarity either. In his well-known Malayalam novel *Chemmeen*, Thakazhi has this in the words of one of his characters: "*Arayar* are of five castes (*jatis)*: Arayan, Valakkāran,

Mukkuvan, Marakkān, then a panchamajāti (fifth caste). Besides these, there is the *Vēlan* in the interior areas."[1] The term Mukkuvan is used here as a subcaste of the Arayan fishing community with no religious connotation. The context of Thakazhi is that of Alappuzha, 150 km north of Thiruvananthapuram. Another novelist, Thikkotiyan, in his novel *Chuvanna Kadal*, uses the terms *Mukkuvan*, *Arayan*, and *Marakkān* interchangeably.[2] His context is north Kerala, some 450 km north of Thiruvananthapuram. These indicate the absence of a homogenous understanding of the term Mukkuvar in popular understanding.

In spite of the fusion of subcastes or elimination of caste names in the course of history, the term Mukkuvar, it appears, was being used originally as a generic term to refer to fisherfolk in general. Ma Huan, an envoy of the Emperor of China, who visited the Kerala coast in 1409, has this to say: "There are five classes of men... the fifth the Mukkuvar, the lowest and the poorest of all."[3] In a footnote, Nagam Aiya clarifies that Mukkuvar refers to the fishermen on the seacoast. The premise that Mukkuvar meant a generic term in early times is further supported by the fact that the name given to the association of fishermen that was formed in Malabar in 1920 was *Akhila Kerala Mukkuva Sangam*. The formation of *Mukkuva Maha Sabha* in Kozhikode in 1933 also supports the same view. Both these were formed for Hindu fishermen.[4] Some suggest that the term Mukkuvar is connected with the Kannada word *Moger*; both the words come from the same root, which means to dive.[5]

Authors generally refer to three main subgroups among the Hindu fishermen of Kerala: *Mukkuvar*, *Arayar*, and *Vālar*. The names of the subgroups may change slightly as used by different authors or used in different places. Velupillai includes *Nuḷaya* as a subcaste in Thiruvananthapuram region.[6] Thurston describes the *Mukkuvans* (sic) as the "sea fishermen of the Malabar Coast," and quotes Buchanan: "The Mucua, or in the plural Mucuar, are a tribe who live near the sea-coast of Malayala (ancient term for Kerala), to the inland parts that they seldom go, and beyond its limits anyway they rarely venture. Their proper business is that of fishermen, as palanquin bearers for persons of low birth, or of no caste; but they serve

1 Thakazhi Sivasankarapilla, *Chemmeen* (Kottayam: D.C. Books, 1995), 33.

2 Thikkotiyan, *Chuvanna Kadal* (Trichur: Mangalodayam), 1963.

3 Nagam Aiya, *State Manual*, Vol. I, 280.

4 Udayabhanu, *Dhivara Community*, 96.

5 Kalpana Ram, *Mukkuvar Women*, 1.

6 Velupillai, *The Travancore State Manual*, Vol. I (Trivandrum: 1931), 401.

also as boatmen."[7] He also refers to the Gazetteer of Malabar 1908 that speaks of the *Mukkuvans (sic)* thus:

> A caste, which according to a probably erroneous tradition came originally from Ceylon [today's Sri Lanka], is that of the Mukkuvans, a caste of fishermen following *marumakkattāyam* (inheritance through the female line) in the north, and *makkattāyam* (inheritance through the male line) in the south. Their traditional occupations also include *chuṇām* (lime) making and *manchal* bearing (a *manchal* is a kind of hammock slung on a pole, and carried by four men, two at each end). In the extreme south of the district, they are called *Arayar*, a term elsewhere used as a title of the headmen. North of Cannanore there are some fishermen, known as *Mugavars* or *Mugayars*, who are presumably the same as the *Mugayars* of South Canara.[8]

However, Thurston adds a footnote: "I am informed that the Mukkuvans claim to be a caste distinct from the *Arayans*."[9]

Thurston remarks that the Mukkuva caste has supplied many converts to the ranks of Mohammedanism. Quoting the Madras Census Report 1891, he says, "Conversion to Islam is common among this caste. The converts are called *Puislam* or *Putiya* Islam (new Islam)."[10] Thurston's familiarity has been more with the Malabar Coast than with the Travancore coast; he makes no mention of Mukkuva converts to Christianity at all.

Some studies refer to the Arayar and the Mukkuvar as two distinct groups of fishermen in Kerala. Lamercinier discusses both groups among the lower castes and the outcasts. "The Arayas," she writes, "were not exclusively fishermen. There were among them both farming classes and others who lived in the mountains."[11] "The Mukkuvans," she adds, "were another fishing clan, a large number of whom were employed in the pearl fisheries. On the religious side, they had been influenced by Buddhism and Jainism.... On account of their social position outside the caste structure, a large number of them converted to Christianity in the seventeenth century—and many also to Islam."[12] The distinction she makes between Arayar and Mukkuvar as belonging to two different clans is a doubtful proposition. There is also anachronism and factual error in the above account; the pearl fishers of the Coromandel were mainly

---

7 Edgar Thurston and K. Rangachari, *Castes and Tribes of Southern India*, Vol. V (New Delhi: Asian Educational Services, 1987), 106.

8 Thurston and Rangachari, *Castes and Tribes*, 107.

9 Thurston and Rangachari, *Castes and Tribes*, 107.

10 Thurston and Rangachari, *Castes and Tribes*, 111.

11 Genevieve Lamercinier, *Religion and Ideology in Kerala* (Louvain: C.R.S.R., 1983), 262.

12 Lamercinier, *Religion and Identity*, 262.

Paravas, not Mukkuvar; the latter is found rarely along the east coast of India. Besides, the mass conversion of the Paravas to Christianity took place in the sixteenth century, 1536–1537 to be exact, and not in the seventeenth century. Authors also point out that the *Mala Arayās*, one of the hill tribes in the Western Ghats, "have nothing in common with the ... *Mukkuvan Arayan* of the coastal regions. The coincidence of the name is an accident."[13]

From the above discussion, we can deduce that there exists no uniform understanding of the term Mukkuvar. Originally used more as a generic term to mean all marine fishermen, it came to be employed to refer to one particular subgroup or another in course of time. In the specific context of the present study, the term Mukkuvar would mean exclusively the community of marine fisherpeople professing Catholic Christianity.

The Sangam Testimony

The Sangam literature is very valuable in shedding further light onto the Mukkuva past. The coastal zone was given due importance in the literature of the Sangam period and was known as *neithaltiṇai*. The ancient Tamiḷakam, that included present-day Kerala, had been divided into five ecocultural zones (*aintiṇai*), each having its own cultural adaptations. Three anthologies (*Pathitṛupathu, Puṛanānūṛu, and Akanānūṛu*) in the *Eṭṭuttogai* give us valuable information about the *Cēra* country (part of present-day Kerala) of the period. *Pathitṛupathu* is wholly devoted to the *Cēra* dynasty and the *Cēra* country. Twenty-seven poets have sung on the *Cēra* country in their poems, and nineteen *Cēra* kings have been eulogized by them. In *Akanānūṛu*, 64 of the 400 poems relate to the *Cēras*; of these, four verses specifically deal with *neithaltiṇai*. Similarly, 74 of the 400 poems of *Puṛanānūṛu* relate to the *Cēra* country.[14]

13 A.A.D. Louiz, *Tribes of Kerala* (Delhi: Bharatiya Adimjati Sevak Sangh, 1962), 120.

14 On Sangam literature, see C. Balasubramanian, *A Study of the Literature of the Cēra Country—Upto Eleventh Century A.D.* (Madras: University of Madras, 1980); N. Subramanian, *Sangam Polity—The Administration and Social Life of the Sangam Tamils* (Madurai: Ennes Publications, 1980); P.T. Srinivas Iyengar, *History of the Tamils: From the Earliest Times to 600 A.D.* (New Delhi: Asian Educational Services, 1983), etc. Sangam refers to an association or academy of Tamil poets, established in Tamiḷakam (present-day South India), probably between 500 BCE and 200 CE. It flourished under the patronage of kings and chieftains. The Sangam literature refers to the collection of the literary works of these poets, a large number of which have come down to us intact. The major works include a grammatical work, two epics, and some anthologies.

The term *tiṇai* referred to geographical areas and to specific norms of social living as well as to psychological moods. Scholars consider this as part of a probable Tamil theory that cultural variations were associated with specific ecological adaptations.[15]The people living in the mountain zone (*kuṛiñji*) were identified as hunters (*Kuṛavar*); the people of the forest (*mullai*) as herdsmen (*Āyar*); those of the fertile plains (*marutam*) as cultivators (*Veḷḷāḷar*); those in deserts or arid tracts (*pālai*) as warriors (*Maṛavar*); and people of the coastal zone (*neithal*) are described as Mīnavar or Valainar or Valaipparadavar.[16] Subramanian writes:

> The Paradavar or the Mīnavar were the fishermen; they were also called the *Valaiyar*. They owned the small fishing boats called *timiḷ* and they plied a brisk trade in fish. They resided on the seashore... . Their net (*valai*) was supplemented by another large container for holding the catch. Paratha Kumārar mentioned in Śilappadikāram were only fishermen youth.[17]

The society of the Sangam period was essentially tribal in character. Its tribal features are clear from its association with certain totems. Many totemic symbols are mentioned in the literature—either as royal emblems or in the form of a tree or a flower. The Pāndyas had the carp fish (*kāyal*) as the emblem on their flag, probably due to their close link with fish trade and seafaring. Each *tiṇai* had its own characteristic features and its own special deity. Tolkāpiyam (Por. 1.5) enumerates the chief gods of four of the five zones. Māyon, the black one, was the god of the forest land; Cēyon, the red one, god of the hilly land; Vēntan, the god of the agricultural land; and Varuṇan, the god of neithal, the seashore. Varuṇan seems to be the Aryanized form of a preexisting Lord of the Ocean whose name is not recoverable. Some suggest that the name might have been Kadalōn, though that name never occurs in the Sangam literature. In the primitive rituals of fishermen, the horn (or tooth?) of a shark planted on the beach symbolized the Lord of the Sea, a god who has practically disappeared under the Aryan hegemony.[18]

In spite of this general pattern, the culture of the Sangam period seems to have given prominence to the worship of the goddess. The importance given to Koṛṛavai, the war goddess, in the Sangam works is significant. She has other names like Amari, Kumari, Camari, Chūli, Nīli, Cankari, and so on. Later she came to be known as Kāḷī or Durgā. Kumari is a goddess mentioned in the Śilappadikāram; she is said to have an eye on

15 J. Pandian, *Caste, Nationalism and Ethnicity* (London: Sangam Books, 1987), 40.

16 N. Subramanian, *Sangam Polity*, 260–61.

17 N. Subramanian, *Sangam Polity*, 293–94.

18 See P.T. Srinivas Iyengar, *History of the Tamils*, 74–75; Kamil V. Zvelebil, *Tiru Murugan* (Madras: International Institute of Tamil Studies, 1981), 7–9.

her forehead. There was a belief that a bath in the holy waters of the cape of Kumari could ward off the sins of adultery and such other.[19] The sea itself is Kadalamma (Mother Ocean), and the ocean deity used to be depicted as feminine.[20] Worship of a mother goddess, especially of Amman, together with that of Murugan, was popular in ancient Tamiḷakam. These factors point to a Pre-Aryan tradition of goddess worship since the gods of the Aryans were mostly masculine. This is an aspect neithal fishermen seem to share with other maritime peoples elsewhere.[21]

The above discussion indicates that the people of the seashore had evolved a culture of their own that could be called "neithal culture." It also manifests its own characteristic religious beliefs and practices giving shape to what could be called "neithal religion." In spite of the strong influence of religious traditions from northern India—of Vedic Hinduism, Jainism, and Buddhism—this neithal religion appears to have survived and maintained its identity. For, the Mukkuva community was never treated as part of the rigid hierarchical caste structure of the mainstream society in Kerala. That they were pushed to the geographical extremes of the land, making them fringe-dwellers both geographically and socially was analyzed in detail by Kalpana Ram.[22] Neithal was placed at the lower stratum of the *tiṇai* structure, just above the Pālai, as if almost excluded from civilization. Literally and symbolically, the coastal fisherfolk, the Mukkuvar, "live on the margins" of the human society, or are "fringe-dwellers," in the terminology of Ram. Though this hierarchical scheme is contested by some authors, the point is a valid one.[23] John Kurien uses the concept *outlier* to refer to the same reality.[24]

The Christian Religious Roots

To my question "Can we say that your forefathers were Nuḷayas (or Arayās) before joining the Christian fold?" most Mukkuva respondents gave me a categorical "no" for an answer. Denying any such link with the

---

19 N. Subramanian, *Sangam Polity*, 373.

20 Lamercinier, *Religion and Ideology*, 3.

21 The fishermen of north Karnataka (*Mogaveras*) called the sea by the name Ganga Māta, for they got their bread from her. The Taoist goddess *Ma-tsu* is the protector of fishermen and sailors. Sedna, the sea-goddess, was considered the guardian of the marine animals by the Eskimos. See G.R. Krishna, *Caste and Tribes of Fishermen* (New Delhi: Discovery Publishing House, 1990), 171; *Encyclopedia of Religion*, s.v. "Goddess Worship: An Overview," by James J. Preston.

22 Kalpana Ram, *Mukkuvar Women*, 4–7.

23 N. Subramanian, *Sangam Polity*, 180.

24 See his monograph, "The Kerala Model—Its Central Tendency and the Outlier" (Trivandrum: International Congress on Kerala Studies, 1994).

Nuḷaya community, they would say, "We have been Christians; Nuḷayas are Hindus". But the more informed among them agree that the "pagan" fishermen who were converted by Francis Xavier and his companions, in all probability, might have been Nuḷayas. Historical records show the link in a better light. There existed an organization called *Thekkan Thirvithāmkuṛ Kathōlika Araśar Sangam* that was started in the 1920s.[25] Velupillai also speaks of "Catholic Araśar" to refer to the Catholic community of fishers. He also speaks of the Latin Catholic Community as composed of "Mukkuvar or Arasar, Paravas or Bharathas."[26] The testimonies by the Hindu Nuḷaya residents of Panathura confirm the Nuḷaya ancestry of the Catholic fishermen. The reasons for their refusal to own up their Nuḷaya or Araya past need to be searched against the background of their Christian formation by the missionaries.

The missionary documents, though dealing mainly with the Christian conversion movement among the marine fisherfolk, are valuable sources of information for this inquiry. There are also some passing references in the "Travancore Mathilakam Grantavari."[27] Marthanda Varma Maharaja, who reigned from 1729 to 1758, mercilessly killed all the male members of the families who joined in the plot to assassinate him in 1736. All their women and children were either exiled or sold to the fishermen.[28] It is not certain whether this incident refers to the Christian fishermen or not; that a major part of the Travancore coast was Christian by then, is beyond doubt.

The Christian history of the fishing communities starts only after the arrival of the Portuguese. Discovering a new sea route to India, Vasco da Gama reached Kozhikode in northern Kerala on May 21, 1498, and a second time in October 1502. His first trip was followed by another expedition by Pedro Alvarez Cabral who arrived at Kozhikode on August 30, 1500. Eight Franciscan friars had accompanied Cabral. The Portuguese in the early stages confined themselves to setting up factories at Kannur and Kochi and competing with the Arabs in winning the favor of the local kings. At Kochi, Vasco da Gama was visited by the queen of Kollam who expressed her wish to establish trade relations with the

---

25 See *The Latin Christian*, a weekly newspaper published from Valiathura, August 16, 1934, 1.

26 Velupillai, *State Manual*, Vol. I, 434 and 875.

27 These are royal records of the rulers of the Travancore kingdom.

28 A.P. Ibrahim Kunju, *Marthanda Varma: Adhunika Thiruvithāmkūrinte Udayam* (Thiruvananthapuram: Government of Kerala, 1990), 30; Nagam Aiya also mentions this incident in his *State Manual*, Vol. II, 475, though he, by mistake, mentions 1773 as the year.

Portuguese; with this, the Portuguese presence reached up to Kollam.[29] Afonso de Albuquerque who came to India in 1503 made a peace treaty with Kollam, and thus a factory was set up there. (Albuquerque came a second time to India in 1509 as the second Governor). Albuquerque left a priest, Fr. Rodriego, O.P., at Kollam to instruct the traditional (St. Thomas) Christians of the place and to convert the non-Christians.

Later, every Portuguese fleet came to India with a group of priests, both religious and diocesan, initially to attend to the spiritual needs of the Portuguese, but later to do evangelization work among the Indian people. The first instance of conversion of a Hindu occurred at Kozhikode in 1500 as a result of the work of the Franciscans who came with Cabral. But the work could not be continued due to the Portuguese conflict with the Samuthiri (*Zamorin*, as seen in colonial documents), the ruler of Kozhikode. More systematic evangelization work was started at Kochi, and a Franciscan Friary was erected there in 1523. In 1510, an Arel (*Arayan*), one of the chief port officers of the Raja, with his whole family and his clan numbering over a thousand, got converted and became Christians. The Arel was christened Antonio Real. This incident encouraged the Arayar of Porakad, Kayamkulam, and Kozhikode to think of embracing Christianity; but no records of its outcome are available. According to a letter of the Vicar of Kochi in 1514, there was a good number of both Īzhavar and Mukkuvar among the new converts.[30] But the opposition by the Raja of Kochi did not allow the conversion work to progress. Albuquerque even tried to convert the Raja, but to no avail. However, the Portuguese built their first church in India at Kochi, and the Vicar of Kochi became, in course of time, the Vicar General, having jurisdiction over all the vicars of other fortresses and clerics in India.

In Kannur too, some evangelization work took place in the early period. The first converts were twenty children, who were left orphans after the massacre by Vasco de Gama in the ship of the Sultan in 1502. Later, the Portuguese were trying to convert men and women of lower classes, and occasionally some Nayars, to Christianity. This was resented by the Raja, and he wrote to the Portuguese king in 1507:

> It is my desire that certain sections of the people in my kingdom whom I and my *nayars* hold as slaves and who belong to two castes, viz. *tines* (*tiyans*) and *mucuas*, may not be converted to Christianity; nor should *nayars* and *brahmins*. The conversion of the slaves will give rise to

---

29 Mathias Mundadan, *History of Christianity in India*, Vol. I (Bangalore: Theological Publications in India, 1984), 265.

30 Mundadan, *History of Christianity*, 359.

> conflicts between them and my vassals. The *nayars* derive their income from them and they do not want to lose it.[31]

The letter of the Raja prompted the Governor to impose restrictions on the conversion of the slaves from the Hindu and Muslim communities. However, later records show that in 1514 there were 344 Indian Christians at Kannur; 85 were converts from Islam, 7 were Nayar converts, 160 came from the lower castes like Tiyans and Mucuas, and the rest were children, 33 of whom came from the same low castes.[32] In Kozhikode, the Portuguese civil authorities did not positively encourage proselytization, respecting the religious sentiments of the Samuthiri and his subjects.

In order to understand the Catholic background of the *Mukkuva* fisherfolk of Travancore, it is necessary to have a glimpse of the mass conversion of the Parava fishermen along the eastern coast of Tamil Nadu in 1536–1537. This event was acclaimed a landmark in the history of Latin Christianity in India and had an immediate impact on other communities especially the Mukkuvar, the immediate neighbors of the Paravas along the coast.

The Pearl Fisheries, which were about 80 km of seashore, extended from Kanyakumari to Rameswaram along the east coast. In the sixteenth century, it had some twenty-two fishing villages. Parava fishermen whose occupation included pearl-diving besides fishing inhabited this strip. Part of the fishery coast in the sixteenth century came under the king of Kollam, and the rest was part of the Vijayanagar Empire. The lucrative pearl trade naturally attracted the neighboring rulers besides the Muslim Arabs who had the monopoly over the seaborne trade. In the process, the Paravas came to be badly exploited by the Arabs with the support of the *Pāndyan* kings under whose jurisdiction the pearl fisheries remained. Many were reduced to the state of slaves and day laborers under the Arab traders and the Kāyalar (Muslim) fishermen. The Portuguese also arrived on the scene in 1523 attracted by the pearls; this set off a series of clashes and battles between the Portuguese and the Muslims. This context forms the background for the mass conversion of Paravas to Christianity. Some historians speak of a quarrel between the Paravas and the Muslims over the issue of an assault on a Parava woman. A mutiny of the Paravas followed, and that was with royal support. The head of a Parava was rewarded five *paṇams* and the Paravas were forced to go into hiding.[33]

31 Mundadan, *History of Christianity*, quoted on 379.

32 Mundadan, *History of Christianity*, 380.

33 P. Thomas, *Christians and Christianity in India and Pakistan* (London: George Allen and Unwin, 1954), 52. *Panam* was the currency of ancient Kerala, either in gold or in silver.

The name of Joao da Cruz, a *Choṭṭi* convert of Kozhikode, is found closely associated with the Parava conversions. Being sent to Lisbon as the ambassador of the Samuthiri in 1513, he came to embrace Christianity. On his return, he was dismissed from service on account of this. Later, he got into horse trade that landed him in Kanyakumari selling horses to the "Great King" of Kollam in 1535. While waiting there for his payments, he was contacted by the Parava leaders with a desire to become Christians. When he returned to Kochi, he brought along fifteen Parava leaders (*Paṭṭangāṭis*) and presented them to Pero Vaz, the Captain, and to Miguel Vas, the Vicar General who happened to be at Kochi at that time. The leaders asked for baptism, the services of priests to baptize their people, and Portuguese help to protect them. The Captain was dubious about the matter because of their small number; so one of the Paravas left Kochi and returned with seventy more. With the Vicar General acting as an intermediary, they all received baptism. George Schurhammer writes:

> At the Pearl Fishery in March and April [1536], the mass baptisms of the men took place without any further, more detailed instructions, since no one knew their language. This was followed by the baptism of the women, the elderly, and the children, who had remained behind in the villages. The following year, during the next pearl fishery, there were further mass baptisms of the Paravas who lived in the north on the other side of the Tambraparni River in the territory of the king of Vijayanagar in Tuticorin, Vembar, and Vippar. By the end of 1537, the entire caste had accepted Christianity, and further conversions were anticipated. When Cruz returned from the fishery coast this same year, the king of Travancore promised him that he would allow the fishers of his territory to become Christians if the Portuguese governor would again grant him permission to buy horses, which he needed for the war of the Great King with Vijayanagar.[34]

With the battle of Vedala in 1538, which saw the end of Muslim naval power, supremacy over the Indian coast fell into the hands of the Portuguese. As a result, the Paravas were greatly freed from the oppression they were subjected to.

Francis Xavier Enters the Scene

In 1542, Francis Xavier arrived in India together with Martin Afonso de Souza, the Governor (1542–1548). He was briefed by Miguel Vaz, the Vicar General, about the Parava Christians, and Xavier reached the fishery coast in October 1542. Three Parava students were nearing completion of their priestly training at Goa, and they were asked to accompany Xavier to the fishery coast as interpreters. Xavier immersed himself in

[34] Schurhammer, *Francis Xavier*, 263.

instructing and organizing the Christian communities of the Paravas since little instruction was available to them after their conversion.

However, Francis Xavier is more closely associated with the conversion of the Mukkuva fishermen of the Travancore coast, the western coast extending north of Kanyakumari (Cape Comorin). This area was predominantly being inhabited by Mukkuva fishermen interspersed with Paravas, especially close to the southern tip. The Travancore coast, for Xavier, meant the stretch from Cape Viḷiñjam (Vizhinjam) where the territory of the Raja of Travancore started, up to Cape Comorin.[35] Xavier seems to have entered into secret negotiations with the *Pāndya* Kings and the Great King of Kollam, both of whom were preparing for war and were seeking Portuguese support.[36] The Tiruvadi Rajas, Unni Kerala Tiruvadi, the ruler of Kollam, Nanjinad, and southern Tirunelveli, and his brother Marthanda Varma who was ruling Travancore were fighting against the *Pāndyas* in 1544. Both sides were eager to get the favor of the Portuguese Governor Martin Afonso De Souza. When the negotiations with the *Pulās* (nobles) of Tirunelveli supporting the *Pāndyas* failed, Afonso asked Xavier to visit the Tiruvadis, offering Portuguese support.

> Xavier met Rama Varma and his brother Marthanda Varma at Tiruvankod, the capital of the latter. The two princes were very thankful to Xavier for all the trouble he had taken for their sake. They now had it announced in the fishing villages of Travancore that the people should obey Xavier as they would obey the Raja himself and that all those who wished to become Christians could freely do so. Rama Varma gave Xavier two thousand *fanams* for the erection of churches.[37]

One legend has it that Xavier stopped the Badaga army of the Nāyak of Madurai marching against the king of Travancore at Mount Comorin (probably at Aṛuvāymoli pass). Wearing a black robe, he went to face the army and rebuked them. Not able to stand his fiery look, they turned around. On his way further, Xavier met the king of Travancore marching with his army against the Badagas. After hearing what had happened, the king embraced Xavier and said: "I am called the Great King, and from now on you will be called the Great Father."[38] It may be noted that Xavier was everywhere spoken of as the "Periya Padre" in the local language that means the "Great Father," from Manar in the north to Travancore in the south.

---

35 Schurhammer, *Francis Xavier*, 292.

36 Schurhammer, *Francis Xavier*, 433; see also Joseph Thekkedath, *History of Christianity in India*, Vol. II (Catholic History Association, India, 1982), 184.

37 Schurhammer, *Francis Xavier*, 460–63.

38 For more details, especially on related historical issues, see Schurhammer, *Francis Xavier*, 443.

Schurhammer writes that the reputation of the *Periya Padre* had extended beyond the boundaries of the fishery coast, and that "the Mucua fishers of Manakkudi on the other side of Cape Comorin repeatedly asked him to come and baptize them and their kin in the villages of Travancore."[39] The Mukkuva fishers then lived in fourteen villages, twelve in the territory of Marthanda Varma (Travancore), and two at the extreme south, in the territory of the Great King of Kollam. Records show that Xavier converted nine or ten "pagan" villages from Vizhinjam to Perumanal.[40] Schurhammer quotes H. Henriques (1558) that the inhabitants of two villages in Travancore had been Muslims before they were baptized by Xavier, and suggests that probably these were the two, which lapsed back into Islam at the end of 1552 under the influence of the Moors. There were about 100 Muslims in Vizhinjam in 1615 besides some in Colachel and Thengapattanam.[41]

Xavier began his baptizing work at Poovar, the linguistic and political boundary of present-day Kerala, in November 1544. He got the help of two Indians from Kollam to be his companions: Christovao Fernandes, and "the Malabar Antonio Fernandes" who knew the language, land, and the people of the area. They moved from village to village, instructing and baptizing. Xavier claims to have baptized more than 10,000 people within a month as he wrote in his letter to his companions in Rome.[42] He could cover only thirteen of the fourteen villages, and had to depart suddenly leaving out Manakudi. From 1546 onwards, there was only Fr. Francis Henriques S.J. to attend to the Mukkuva Christians, at times helped by a lay brother.

In 1547, the trend began to change. The Raja of Travancore summoned Fr. Henriques and forbade him from making any more converts in his kingdom. This was to express his annoyance at the Portuguese and the missionaries for not supporting him in his war with the forces of Vijayanagar. Later, the king expelled Br. Balthasar Nunes, S.J., the only companion of Henriques, from his kingdom. The ill-treatment by the Raja and his officials made the Christians suffer much. However, documents show that the Raja opened his kingdom again to the missionaries in 1568. A report of 1568 says that there were seventeen churches and about 8000 Christians along the Travancore coast.[43] In 1601, the number of churches went up to twenty-three, and these were looked after by two Jesuit fathers

39 Schurhammer, *Francis Xavier*, 460.

40 Schurhammer, *Francis Xavier*, 467–69.

41 Schurhammer, *Francis Xavier*, 467.

42 This was dated January 27, 1945; see Henry J. Coleridge, *The Life and Letters of St. Francis Xavier* (London: 1881), 117.

43 See Thekkedath, *History of Christianity*, 188.

staying at Colachel. There were 550 adult baptisms in 1602–1603.[44] Till 1603, Jesuit work in this area remained exclusively along the coast. "Our information regarding the Christians in this region becomes very scanty for the rest of this century," writes Thekkedath.[45] According to a report about the Jesuit mission in India by Fr. Andrew Lopez, S.J., Vizhinjam had 200 Christians in 1644 and the church was under the patronage of Our Lady of the Assumption.[46] A report of 1681 expressly states that there were very few conversions.[47] The Mavelikara pact between the Dutch Company and Marthanda Varma in 1743 mentions, under clause 17, the Dutch right to banish from Travancore all Roman priests of European origin. How this was implemented is not certain. However, the Jesuit work in Travancore might have come to an end with the suppression of the Jesuit Society, first in Portugal, and later globally in 1773. Goan clergy continued the pastoral work along the Travancore coast, but documents of this period remain scanty.

Xavier's own writings give a good picture of the systematic procedure he followed among the Mukkuvar.[48] Usually, he went to a village at the invitation of the people there who desired to become Christians. After reaching a village Xavier gathered all men and boys together; then he explained to them the chief articles of the Christian faith in the local language. He taught them the Sign of the Cross, Confiteor, the Creed, the Commandments, Our Father, Hail Mary, and Salve Regina. He made them repeat what he had said from time to time. This was followed by an explanation of the Creed and of the Commandments. Then, the people together asked God's pardon for the sins of their past life. He recited the articles of the Creed one by one and made them reply that they believed in every article of the Creed. After this, he baptized them with a simple formula. Each one received a palm leaf strip with one's name written in the local language. The men went home and sent their wives and others at home; the same procedure is again repeated and they too received baptism.

Xavier's presumption was that the Mukkuvar (*pagans*) were idolaters and worshippers of the devil, and so he insisted on the destruction of

---

44 Thekkedath, *History of Christianity*, 191.

45 Thekkedath, *History of Christianity*, 193.

46 K.J. John, *Christian Heritage of Kerala* (Cochin: Grand Chevalier L.M. Pylee Felicitation Volume, 1981), 375.

47 Thekkedath, *History of Christianity*, 194.

48 See Xavier's "letter to his companions living in Rome", January 27, 1545 (Coleridge, *Life and Letters*), 116–118, and "Instructions for those of the Society who are on the Fishery Coast and in Travancore", February 1548, see also Thekkedath, *History of Christianity*, 195–196.

their idols and their temples. "When I have finished baptizing the people, I order them to destroy the huts in which they keep their idols; and I have them break the statues of their idols into tiny pieces since they are now Christians."[49] Children and young boys played a major role in this process, as was demonstrated earlier among the Pearl Fishers.

> They [the boys] reproach their fathers and mothers when they see them worshipping idols, and they denounce them by coming to tell me about it. When they tell me about idolatries that are being practised outside the villages, I collect all the boys of the village and go with them to the place where the idols have been erected; and the devil is more dishonored by the boys whom I take there than he was honored by their fathers and relatives when they made and worshipped them, for the boys take the idols and smash them to bits. They then spit upon them and trample them under their feet; and after this, they do other things, which, though it is better not to mention them by name, are a credit to the boys.[50]

Then Xavier moved on to the next village where the same procedure would be repeated.

Valignano reports that if the newly baptized failed to perform their religious duties, or refused to obey the missionaries, or lapsed back into idolatry, the captains of Kollam could exert a certain pressure on them by forbidding them from fishing or by confiscating their boats and catamarans.[51] But very little of such sanctions seems to have been applied.

These narrations illustrate that in the Mukkuva consciousness, no contradictions between their traditional faith practices and the new religious identity were present. Attempts were made to erase their traditional forms of worship and rituals by zealous missionaries and to replace them by the catechism of the Catholic faith. Documents give descriptions of the way catechism classes were conducted in the coastal villages.[52] Xavier had composed *Doctrina Christiana*, the short Portuguese catechism, in Goa in 1542 (that was printed later in 1557); this seems to have been the foundation for his catechetical ministry. He had got the prayers translated into Tamil at Tuticorin in 1544, and also composed *An Explanation of the Apostle's Creed* in Portuguese in 1546 which too was translated into Tamil. He made numerous copies of the latter to be distributed in churches and also ordered this as well as the *Doctrina Christiana* to be taught in the villages. With the setting up of a printing press in Goa in

49 Schurhammer, *Francis Xavier*, 117–118.

50 Schurhammer, *Francis Xavier*, 66.

51 Schurhammer, *Francis Xavier*, 465.

52 Thekkedath, *History of Christianity*, 187.

1557, these were reprinted many times and were in wide use till the twentieth century.[53]

The ecclesiastical structures and offices among the new Christians of the Travancore coast are not as well documented as those on the Coromandel coast. Xavier had appointed *kaṇakkapiḷḷas* in the Mukkuva villages of Travancore, as his own letter testifies, and instructed Mansilhas to set up a school for the children in each village with a teacher to instruct them. He also refers to the *meirinho* (warden or Portuguese law officer), who used to accompany him. The new structure of ecclesiastical officials introduced by Xavier among the Paravas is likely to have been used by him as the model among the Mukkuvar too. Examples are the *modom* (overseer of ecclesiastical details), the *ubedesiyār* (sacristan), the *kaṇakkapiḷḷa* (catechist), and the *vāthiyar* (teacher) besides the office of the vicar.[54] The Jesuits had introduced translation or composition of Christian prayers in Tamil and printed them in book form for wider use from the early stages. *Tambirān Vaṇakkam*, a booklet of prayers prepared by Fr. Henriques in 1577, is considered the earliest printed work in Tamil. *Kristhiyani Vaṇakkam*, another prayer book printed in 1579 was also prepared by him. His later work, *Flos sanctorum a librodo las vidade algunos sanctos transladas en lingua Malabar*, printed in 1586, was a massive work of reflections on the principal mysteries and feasts and the lives of the principal saints of the Church's calendar.[55] It is likely that these works were available to the Catholics of the Travancore coast. The prayers recited by the elderly in the Malayalam speaking Mukkuva areas are still in the Tamil language.

Festivals and pious associations were the other two areas the Jesuits paid attention to. Shrines and festivals were introduced to supplant and divert the Parava attachment to the Hindu festivals at Madurai and Tiruchendur.[56] In 1582, the Jesuits imported from Manila to the church of Our Lady of Pity at Tuticorin a statue known as *Senhara das Nevis* (Our Lady of Snows=*Pani Maya Māta* in Tamil). The festival of this *Paratha Māta* (Mother of the Paravas) that is celebrated on August 5 became the most popular feast along the fishery coast. Roche gives a detailed description, based on ancient records, of the way the feast was celebrated in 1600 with processions, floats, and banners.[57] Later, a novena

---

53 Coleridge, *Life and Letters*, 150.

54 Roche, *Fishermen of the Coromandel*, 67.

55 Roche, *Fishermen of the Coromandel*, 51.

56 Tiruchendur in Tirunelveli district of Tamilnadu was a seaside town. Its *Vaisakham* festival was very famous and attracted people from all over from very early times.

57 Roche, *Fishermen of the Coromandel*, 53.

also was started in connection with the festival. The Confraternity of Charity is the earliest form of pious association we come across, and it was started by Fr. Henriques in 1575 at Punnaikayal. Later the Sodality of our Lady was formed in Tuticorin in 1648; soon it spread to other fishing villages too. Some reports say that the Fathers had appointed a judge and a policeman in every Christian village. Besides, there was also a judge for the whole coast to whom appeals could be made. More serious cases could be referred to the Fathers.[58] It is not certain whether this system existed in the Christian villages of the Travancore coast.

From Neithal Religiosity to a Creedal Religion

The coastal fisherfolk always lived on the fringes of the caste society in Kerala, and never became part of the caste system that determined people's identity and occupational status. At the same time, the occupation of fishing made them marginalized in the mainstream society. Caste-wise, the Mukkuvar were treated as slightly lower than their neighbors—the Paravas and the Shānars. Therefore, their moving over into a Christian religious fold might not have been a heartbreaking experience. They had very little to lose in terms of social identity or status as a result of their religious conversion.

It is shocking to read some of the missionary reports of the period to see how they perceived the Mukkuvar as Christians. Many reports say that the Christian fisherfolk along the coast between Kollam and Cape Comorin was "on the whole poorly instructed in the faith and hard to deal with." There are also references to "the quarrelsome nature of these people."[59] It was a fact that there were not enough priests to instruct them. Valignano, the Jesuit visitor to the coast in 1575, wrote that the people "were living almost like pagans."[60] This statement stands in clear contrast to his observations on the Paravas of the fishery coast who were baptized only a few years earlier. About the Paravas, he said that many of them lived exemplary lives and that it was the best Christian community under the care of the Jesuits in India. The same letter also said that the children there knew the catechism better than the children in many parts of Europe.

The impressions of Fr. Antony Monserrate, S.J., who spent some time on this coast are also revealing. According to him, "although there are some good Christians on this coast, the great majority could not be considered good. Some of them gave much trouble to these fathers." He cites

58 Thekkedath, *History of Christianity*, 189.

59 Thekkedath, *History of Christianity*, 186.

60 Thekkedath, *History of Christianity*, 190.

Image 3. Faith and life unite on the seashore

the example of Fr. Andrew Hernandez who was beaten up "because he ordered the demolition of a little temple that some bad Christians had set up in front of a church in order to obtain the favor of the devil for their fishing." Again, "the reason for their not being good was partly the fact that they were very rustic and given to much drinking."[61] The annual report of 1644 stated that "the Christians of this coast were still not very obedient to the Church, and that they went to the extent of bribing the Hindu officials of the Raja in order to be able to resist with impunity the authority of the priests."[62]

The question of whether the missionaries were justified in painting such a negative picture of these people is a valid one. In their urge for bringing them up in the Christian tradition, the missionaries might have failed to grasp or appreciate the cultural ethos of these people. More relevant, however, are the undercurrents of conflict that characterized the changed religious identity of the Mukkuva Christians. On the one hand, they seem to have embraced Christianity willingly and without coercion as it brought them a new social identity. On the other hand, they seem to have refused to submit fully to the new authority structure and its seemingly "unreasonable" demands. This complex situation, from the Mukkuva point of view, can be seen as an expression of the incongruity that existed between the new and the old in some core areas. This incongruity deserves further probing.

Christianity did not get rooted among the Mukkuvar as in a cultural/religious vacuum; they already had inherited through generations what could be called a neithal religiosity. The Portuguese missionaries who came with a worldview of feudal Catholicism (in which the Catholic religion was the dominant force, and sectors like politics and economics

[61] Thekkedath, *History of Christianity*, 190.

[62] Thekkedath, *History of Christianity*, 194.

were kept subordinate to that) that was prevalent on the Iberian Peninsula found these people "pagans" and idolatrous. This led the missionaries to systematically eliminate the elements of neithal religiosity and to replace them with a creedal religion in the form of Catholicism. The dynamics and consequences of such a displacement are yet to be explored. The conversion experience had a different frame of meaning if seen through the optics of these Catholic fisherfolk.

For the Mukkuvar, becoming part of the Christian fold was relatively a smooth experience unlike in other similar situations. They did not embrace Christianity in order to escape a state of oppression unlike in the case of the Paravas of the Coromandel Coast. Nor did it imply physical coercion or violent conflicts as had happened in many Portuguese settlements in India. In Goa, an aggressive phase of brute force preceded the mass conversions, mostly a political necessity, in the mid-sixteenth century.[63] Religious change in Travancore was relatively free from the use of political power or military action. And that ensured certain continuity with the belief system and practices that existed in the coastal communities. There was no need to break away from the caste hierarchy as the Mukkuvar were never an integral part of it. Further, it is possible that some aspects of common wavelength in beliefs or worldviews might have made the transition a smooth one. Yet, beyond this smooth process, there remained an area of discrepancy and destabilization that continues to disturb the community. It is probably related to the puzzling identity consciousness that the community has come to inherit. The acceptance of a creedal religion meant that the open religious boundary line of their neithal religion was sacrificed and got replaced by a religion with apparently closed boundary lines. Thus, the Catholic fisherfolk became well demarcated and enclosed as never before, vis-à-vis their neighboring coastal people.

It would appear that the Catholic fisherfolk felt uncomfortable with the weight of ecclesiastical structures and authority patterns that were imposed by the Iberian Catholicism. The constant complaint by the missionaries about their "rebelliousness" is partly explained by this fact. In spite of the acceptance of all external signs of the Catholic faith, their basic belief system rooted in the neithal religiosity seems to have persisted. In short, the incongruity between what is officially professed and what is really lived would characterize the present-day religion of these coastal Catholics. Their lived religion is intertwined with this radical

---

63 See Rowena Robinson, "The Cross: Contestation and Transformation of a Religious Symbol in Southern Goa," (*Economic and Political Weekly* 29, no.3 (January 15, 1994): 94 – 102.

incongruity. The historical overview has given us an insight into this paradox, but our concern is more with the present and the future rather than the past. In the next chapters, we shall further explore how coastal Catholics live their religion and how coastal Catholicism sustains them in a world characterized by danger and disaster.

CHAPTER 4

NEITHAL COSMOLOGY: WORLDVIEWS IN CONFLICT?

It is only within the specific worldview of a people that their sacred universe becomes comprehensible. The religious representations that embody the sacred universe provide sense and meaning to isolated events and experiences in their everyday life. If taken in isolation, these lose their meaning. Cumulatively, they provide interpretative schemes and recipes for the conduct of individuals and collectives. This explains why consideration of worldview becomes necessary for a study of religious phenomena. Thomas Luckmann defines worldview as a social form of religion that is elementary and nonspecific.[1] History of religion makes it clear that every religious tradition presupposes a particular view of the phenomenal world that permeates every aspect of the belief and practice of that tradition. The specific way of viewing the world is central to the mode of self-understanding and actualization of every community. For coastal people who live in a deep symbiotic relationship with the ocean and the related phenomena, the optic through which they view and interpret the world around them becomes central to any study of their religion. The term "cosmology" is usually employed to designate such a worldview. Cosmology refers to "the specific view or collection of images concerning the universe held in a religious or cultural tradition."[2] In fact,

---

1 Thomas Luckmann, *The Invisible Religion* (London: The Macmillan Company, 1970), chapter 4.

2 *Encyclopedia of Religion,* 1987 ed., s.v. "Cosmology: An Overview". The terms *cosmology* and *worldview* are at times used interchangeably. Some distinguish both. *Worldview* (Weltanschauung) was introduced by Immanuel Kant as a synonym for cosmology. But today, the term worldview is being used more vaguely to refer to the sum of ideas concerning life and world commonly accepted by a group of people. Cosmology is used to denote more consciously systematized images, doctrines, and scientific views concerning the universe. In this book, both the terms are used interchangeably. See also *International Encyclopedia of the Social Sciences*, Vol.16, s.v. "Worldview."

it is unrealistic to expect a systematic cosmology in the context of the marine fisherfolk. A tradition of speculative articulation and documentation has not been part of their heritage. However, a closer look will reveal that a powerful cosmological foundation has been sustaining them and energizing them for generations. We may be able to trace only fragments of their cosmological wisdom, and these fragments may open for us a window into their life-world and their neithal religion.

The Topography of Neithal Cosmology

We often see fishermen sitting on the seashore as if idle but keenly watching the moods and motions of the vast ocean all through the day. As they sit with their backs against the landmass, the vast ocean in front of them engulfs them as if nothing else exists. It is but natural that the ocean rather than the land becomes the primary reality for them coloring their thoughts, attitude, and behavior. Their unique geographical location places them on the "edge" thus orienting their whole life to a water-world.

The ocean is ever changing and unpredictable. It may change its color and rhythm any time due to factors like tidal movement, change of seasons, and occurrence of storms or cyclones. The sea is calm and inviting at times, but ferocious and frightening at other times. The rough monsoon season is marked by risk and danger to the fishermen as well as to their craft and gear. Certain fishing operations like shark fishing require special expertise and stamina for deep-sea fishing and may involve greater danger. The dangerous scene of shark hunting has been brought out most dramatically by Ernest Hemingway in his novel, *The Old Man and the Sea.* The unpredictability of a catch and the eventuality of an empty-handed return is part of the routine for any traditional fisherman. A fisherman has to rely on his daily labor since fish is a highly perishable commodity. Unpredictability and risk seem to characterize the neithal culture.

The neithal man is basically a hunter-gatherer and a nomad. The nomadic features express themselves powerfully in the migratory character of the marine fishermen that takes them to far away shores in search of fishing grounds. Thiruvananthapuram fishermen, known for their expertise, move all along the east coast and the west coast of the Indian peninsula. These expeditions usually take weeks or months at a stretch. We also notice a reverse process of inward migration bringing fishermen from neighboring districts to the Thiruvananthapuram coast. The vibrant fishing harbor of Vizhinjam is a safe landing center for traditional boats in all seasons. Fishermen from other places come with their boats on trucks and may stay on for months. This brings about change in the demography of the area and has sociological implications too. The absence of the menfolk for a long duration every year leaves the leadership of the community in the hands of a few merchants and middlemen. Domestic management role falls on the shoulders of the womenfolk as a natural consequence. Features of a matrilocal tradition are evident in the coastal communities, although it is fading in recent times.

A hunting instinct is intrinsic to the occupation of marine fishing and that marks it as distinctive from the agrarian culture of the mainstream Kerala society. The impact of the neithal context and its culture on the psyche of the fisherfolk deserves serious study. The rugged manners and aggressive dealings of a fisherman will not be easily understood by a person with an agrarian background. The patient waiting of the farmer nursing the seedling and caring for its growth through watering and manuring that lasts many years looking up to the time of harvest in due season is totally alien to the life-rhythm of marine fishermen.[3] The cosmological conceptions of the fisherfolk can be understood only against the specific features of a neithal culture.

Kenkādeviamma and the Phenomenon of the Ocean

The normal term the fisherfolk use to refer to the ocean is *kadal.* It is an ambivalent term and can mean either the ocean in general or the breakers that rise up and hit against the shore in fury. In the monsoon months of June–August, the sea becomes very rough and they call it the "big sea" (*valiyakadal).* During Onam time also the sea becomes quite rough, and it is called *ōṇakkadal.*[4] She is affectionately called Kadalamma (Mother Ocean). It is their mother, the eternal provider. They have an unfailing faith in the goodness and bounty of the ocean. "She will provide for tomorrow also" is the refrain often heard in a fishing village. Why should one save for tomorrow, and for generations to come? While pushing the boat over the waterline, every fisherman can be seen paying homage to Kadalamma reverentially touching her and making a sign of the cross on his forehead. The feeding mother needs to be revered.

There is an element of incongruity too in their attitude to the sea. Though she is a bounteous mother, she is also a place of mystery and danger. Every fishing expedition is a risky adventure. For a fisherman,

3 Raymond Firth, in his study of Malay fishermen, discusses the contrast between an agricultural economy and the economy of a fishing community; see his *Malay Fishermen and their Peasant Economy* (London: R.K.P., 1971), 2–5.

4 Onam is a harvest festival that is celebrated in Kerala in August–September every year. According to legends, it is on this day that Mahabali who had ruled the land in ancient times returns to visit his subjects.

fishing is a struggle with the ocean and your life is ever under threat (*Jīvane paṇayam vachuḷḷa jōli*). "Never go fishing with the assurance that you will get back safe," elders say. To be lost in the midsea for days, with no sun or stars in a cloudy sky, is the ultimate experience of human vulnerability and helplessness. They know that humanly, there is very little that one can do in such situations. "Simply drop your anchor and stay calm for the storm to subside; never try to rush ashore, that will only lead to greater trouble," they advice the young. Instead, they place all trust on those on the shore—the prayers and supplications of the dear ones to the protector saints. Such experience, probably, is what makes these fisherfolk a deeply religious people. "The safety of the fisherman out in the sea is in the hands of the woman on the shore," is the age-old maritime wisdom that has been portrayed in the well-known Malayalam novel, *Chemmeen*.[5]

The primacy of the sea in relation to land is a central feature of neithal understanding of the world. They say it is six and a half parts the ocean, and only half a part the land (*Āṛara thankam kadalum ara thankam karayum*) signifying the relative importance given to the sea. Scientists confirm that only 29 percent of the earth's surface is land; 71percent of it is water. Panidasan, an active fisherman, sounds mystical when he explains it: "It is the sea that is all around us, and it is water that supports the weight of the land. Water is power. The sea existed before the land came to be. In the beginning, that is, before God created the world, there existed only God the Father and the waters!" In his conception, *land* refers not to the landmass of planet earth, but to the immediate, familiar land, that is, the land of his village. The sky is experienced as a roof over the sea, without the support of any pillar. While in the sea, the horizon is experienced as moving in either direction—toward the deep sea far away or toward the mountains on the land. The shoreline *(kararēkha)* is the boundary of the ocean on one side acting as a landmark, while the horizon over the waters helps him to predict rain or storm by reading changes in its glow.

The land and the sea are two inseparable realities for a fisherman. However, he spends a major part of his life in the sea—that being the sole ground of his sustenance and survival. It is there that he does his "planting" and "harvesting" and "storing." This marks him as distinct from most non-fishing people for whom land is the sole basis of sustenance. A fisherman can hardly survive away from the ocean. In this sense, the ocean becomes the primary reality for him.

---

5 This is clearly depicted in the Malayalam novel, *Chemmeen*. See Thakazhi Sivasankarapilla, *Chemmeen* (Kottayam: D.C. Books, 1995).

James, a young fisherman from Vizhinjam, shared the experience of his first fishing venture out into the deep sea as follows:

> It was an experience I can never forget. Far out into the sea there was no trace of land. It was only water on all four sides. I felt I was sitting in a huge bucket of water, surrounded on four sides by a massive wall of blue sky, covering even the top like a canopy... . It was a fantastic experience to be there. But it was also an experience of fearful helplessness.

To be "in" for a fisherman is "to be in the sea." The sea is his world. The shore, not the land, is his point of reference, his *axis mundi*. The land is of little significance; it is only a casual landing center for a recess from his workplace, the ocean. It is significant that he calls the deep sea *uḷkadal* (the interior sea), and the interior land mass *karapuṛam* (the outer land), and the seashore *kadalpuṛam* (the outskirts of the sea). His two reference points—*uḷḷu* and *puram*—stand in contrast to the common conception in the Kerala society where *uḷḷu* is used for designating *uḷḷnādu* (the interior of the land), and *puṛam* refers to *puṛamkadal* (the outer sea). This apparent reversal of directionality in the thinking of fishermen is no linguistic slip but is intrinsic to the neithal worldview.

The coastal rhythm of life is attuned more to the ever-changing features of the ocean, and to the steady waxing and waning of the moon. Therefore, it is natural that an average fisherman feels like an alien on land. Thadeyoos, another young fisherman from Vizhinjam, shares his experience: "After our return from fishing we do not know what to do or where to go; time doesn't seem to pass, and it is real boredom. We never feel that boredom while in the sea; we have one thing or another to do always. Even the constant variation in the sea is worth watching. Every fishing trip is a new experience." Many others share the feeling as that of Thaddeus.

Since land is less important, so too is a house as a dwelling place. A casual observer will notice that many houses are being used more as a storehouse of fishing gear than a residence. Since a fisherman has to stand the sun and the rain and the storm in the open sea, whether day or night, a stable shelter on land is of least priority. The menfolk prefer to sleep at night on the sandy beach, partly for the convenience of launching for work at odd hours. Women and children too frequently sleep outdoors—in the by-lanes in between clustered houses. Ownership or boundary, privacy or property, thus become insignificant. Most of the settlement areas along the beach, except the new colonies, are still the property of the church. Only a tribal configuration will help us understand these features in proper perspective. Today these are undergoing fast changes. Ownership titles and boundary demarcations have become necessary, and the settlements are expanding to the more spacious outer slopes, with compound walls and terraced buildings. However, old tribal modes of conduct and rhythm of life refuse to disappear.

Two phenomena in the sea are very significant to seagoing fishermen. One is a special glow that appears in the seawater (*kavaru*) and the other is the coloration of the ocean surface (*kaṟa*). The seawater displays a glowing effect at night when there is no moonlight. The same glow is visible on the wet sand in the dark when the surf recedes. Fishermen may not be able to explain the reason, but they know from experience that *kavaru* can ensure a good catch. *Kavaru* makes fish glow in water, or rather, the fast movements of fish cause ripples that make fish more visible. In the bright moonlight, *kavaru* vanishes, and fish tend to go down. It is important to distinguish between two types of *kavaru*. One type, known as *kavaruveḷi*, is as bright as a tube light, they say, and so fishing is impossible in it. The second type, known as *adakkamkavaru*, is softer, and would disappear when the water is stirred. This *kavaru* is conducive for fishing. The second phenomenon is the coloration of the ocean surface known as *kaṟa* or *kaṟanīru*. Elders testify that there are seven types of *kaṟa*—red, blue, black, white, etc. The ocean surface may look like a red carpet because of *pūnkaṟa*, but when we go closer it appears green. Its unpleasant odor can cause a vomiting sensation. When there is *pūnkaṟa*, *kavaru* will be visible even in moonlight.

Fishermen explain the phenomenon of coloration in various ways. It is the muck from the bottom of the sea welling up, some say. It is the sperm ejected by male fish, according to some others. But scientists point out that seawater is colorless. It appears colored depending upon various factors like depth, presence of plankton and other microorganisms, and the presence of certain solvents. The ocean would appear green where it is shallow, blue, ash, or deep green in deeper waters. Usually, plankton is the reason for the sea appearing unusually red or green; change in seasons can cause some types of plankton to rapidly multiply through fast reproduction thus changing the color of the ocean surface. Shrimps and crabs also have a plankton stage in their lifecycle. Whatever be the scientific explanations, the ocean for the fisherfolk is a complex world that consists of fish and water, light and darkness, movement and stillness, color, and glow. Its essence—and beauty—is in its ever-changing dynamism and interplay of all these factors. As the fish in the sea adjusts to the endless change and makes it their home, so too, a fisherman finds it "his world." Its *axis mundi* is not on the land but in the sea.

That there is nothing unholy in the sea is a general belief among the fisherfolk. "There is dirt in the sea if there is dirt in one's eye" (*kaṇṇil chavaṛundenkil kadalil chavaṛundu)*, elders say. Her purity is such that no external dirt can pollute her. Hindu fishermen usually use seawater in their purification rites both at home and in the temple believing that it is more holy than cowdung water. Mother Ocean symbolizes purity and sacredness. The "sense of the holy" with which they approach the sea extends to the whole marine life-world. This is evident even in their fishing operations. They know which fish to catch and which one is to be revered. They study the character and behavior of each category of fish and treat them accordingly. They are guided by certain moral codes that are unique to the neithal culture. The whale varieties are usually avoided—not only due to their huge size and potential danger but also because of the sacred myths that are woven around them. The moral code of the sea prohibits them from calling a whale by its popular name, *kadalāna* (sea elephant). Honorific titles like "big fish" *(valiyameen)* or "dear little one" *(chellapiḷḷai)* are used instead. Certain customs prohibit them from uttering the name of any fish before it is caught; or else, it would slip away, they believe. They say a prayer to St. Antony for protection from attacks by big fish. One should never tease a fish while on a fishing operation; it may chase you endlessly, elders say, and they narrate stories of such encounters. Fishermen in many places believe that fish in clusters come close to the shore and jump up in joy at the time of their parish feast.

There are many stories and legends around the crab. One variety wears a cross on its forehead, and so it is called *kuriśuñjandu* (cross-crab). In earlier days, fishermen used to avoid catching it because of the venerable cross it carried. The story behind it was narrated by Markos, an elderly fisherman from Vizhinjam.

> During his mission journeys, one day Francis Xavier called a meeting of the people of the village, but nobody turned up for the meeting. In anger, he turned to the fish of the sea and invited them to the meeting. A crab, usually known for its slow motion, was the first one to arrive. Pleased with it, Xavier drew a cross on its back. The last ones to arrive were the ribbon fish and the catfish, and so Xavier cursed them. That is the reason why these carry a 'stone' on their back.

"Fishes have their own language, but we humans are unable to understand it," affirms Markos. One person who has researched into the psychology and language of the fish is Panidasan, a skilled fisherman from Vizhinjam. He is familiar with the routine of each fish variety, and he is convinced that it is the secret of his getting good catch daily. "The main meals-time of fish is between 3.30 p.m. and 6.30 p.m. It is then that they feel hungry and swallow the bait readily. The peak time for eating is around 5.00 pm. After 6.30, they usually do not eat anything," Panidasan asserts. Based on such knowledge, he had been building up *pāru* (underwater cleft that serves as a breeding ground/shelter for fish) by carrying a couple of boulders in his boat daily while going fishing and depositing them at a marked spot. That spot, unknown to other fishermen, became for him a rich fishing ground in course of time. Today, the scientific community has acknowledged the ecological importance of cultivating and protecting *pāru* on the seabed.

The Astronomical Wisdom of *Kaṇiyam*

Kaṇiyamnokkal is the coastal term for the age-old maritime knowledge and skill that enables the fisherfolk in reading the signs of the sea and the sky in order to calculate the favorable time for fishing and allied operations. At one level, it relies on a system of convergence involving distant objects like mountains, towers, or planetary bodies in locating a point in the ocean and marking it for future use. The sense of direction, the wind and the currents, and the regularity in the movement of heavenly bodies, all have a role in *kaṇiyamnokkal*. *Kaṇiyam*, in short, is the traditional maritime science of computing the space-time configuration in a fishing context that has been handed down through many generations. Unfortunately, much of this astronomical wisdom remains undocumented and is kept alive through oral tradition.

We may hear a fisherman say, "I am going south," (*thekku pōṇu*) meaning he is going to the seashore. This is common usage among fisherfolk along the seashore south of Vizhinjam. This usage is bound to confuse anyone accustomed to the textbook description of Kerala, as having the Arabian Sea (or Lakshadweep Kadal) as its western boundary, if the ocean actually is in the south or in the west. Closer scrutiny using a compass would show that the reading of the fisherfolk is closer to the truth. This is partly because of the geographical position of the coast beyond Vizhinjam where it takes a sharp turn to the east thus bringing the sea to their south. The knowledge of the fisherfolk comes not from textbooks, but from their observation and experience. The linkage between the movements of heavenly bodies with the shoreline helps them determine the directions.

Accordingly,

South: the sea, the world of fish

North: the land beyond, the countryside

East: the coastline toward Kanyakumari and beyond

West: the coastline toward Kollam and beyond

The terms they use to refer to the directions are:

uḷḷē = south, the interior

karē = north, the outlying

kīḷā = east, the lower

mēla = west, the upper

Language scholars say that words referring to directions come into use based on sunrise and sunset, the former referring to the east and the latter to the west. In Tamil and Malayalam, the term *kizhakku* (from *kīzhu*=below) is used to refer to the east; *mēlkku* (from *mēlu*=above) refers to the west. These usages are still seen in old land documents.[6]

This conception of the directions can diagrammatically be shown as given below.

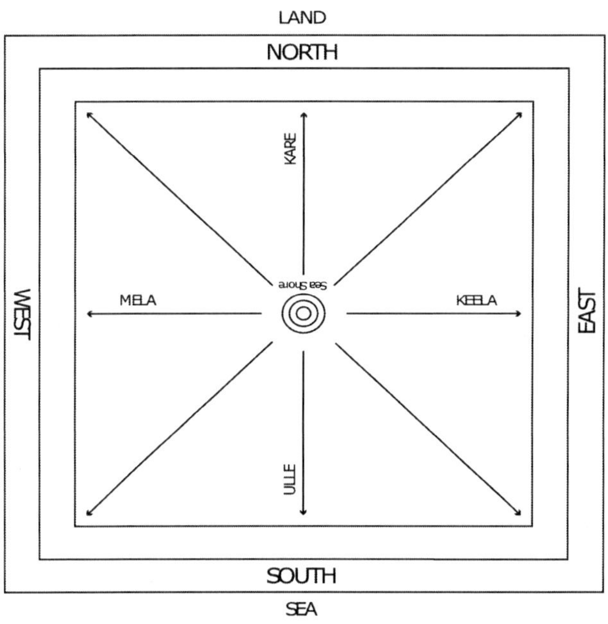

Fig. 2. The Directions

This peculiar sense of direction has much bearing on the occupational world as well as the ritual life of the coastal people. Traditional wisdom indicates that the extension of bodies in space is in three directions—length, breadth, depth, or height. The lines representing these directions point to the four cardinal directions at the horizontal plane besides the

6 K.M. George, in *Kerala Charitram*, Vol. 2, (Malayalam) (Ernakulam: Kerala History Association, 1974), 599–601.

two vertical directions. The four cardinal directions, the four secondary directions that lie between the cardinal directions, and the two vertical directions together constitute ten directional indicators.

Floating on the waters in the open sea, fishermen are very much at the mercy of the wind and the current that vary constantly according to times and seasons. The behavior of fish also gets affected by these changes, thus requiring a change in fishing operations and in the rhythm of coastal life. Fishermen speak of nine types of wind that are vital to their life and work. The diagram below presents them in a nutshell.

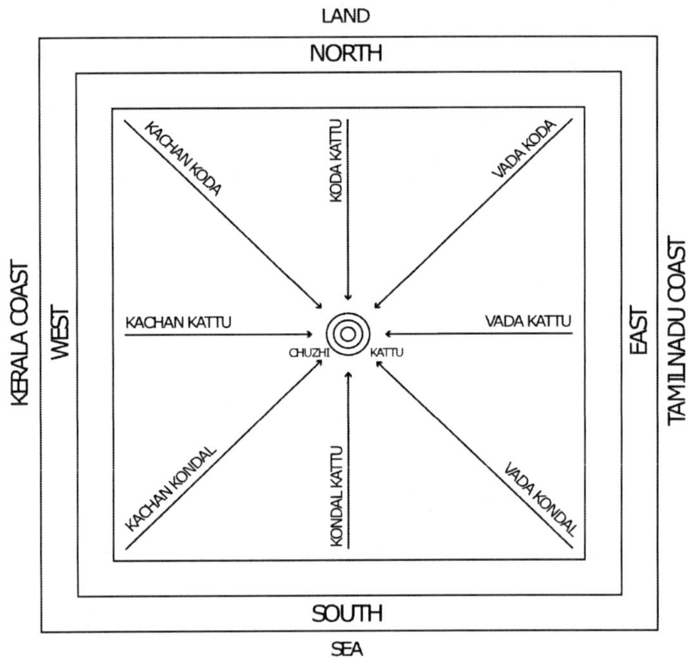

Fig. 3. Winds and their Directions

Depending on the wind speed, they call them either male or female. *Kachān* and *kōda* are said to be female (since these sustain them like a mother, they say). *Kachān* is the most favorable one for fishing. *Vāda* and *kondal* are considered male and are very powerful winds that make fishing operations unsafe. *Kondal* that is exceptionally powerful is called *maṛiakondal*. The currents also behave very similar to the winds. Fishermen say it is the currents that bring in or take away the fish. For, experience has taught them that fish always move against the current, and this knowledge is valuable to their work. “Fish would make a whimper when the sea becomes rough; that is a good time to cast our nets. In still waters fish tend to stay idle at the bottom," says Carlos, a 70-year-old fisherman from Vizhinjam.

Scientists say that ocean currents are caused by factors like the speed and direction of the wind, variation in temperature, and change in the density and salt content of the seawater. Water in the Arabian Sea is saltier than in the Bay of Bengal. Similar to the names given to the winds, fishermen have named nine main categories of currents or pulls. If the pull is to the east, it is *kīḷāṭṭuvalivu*, which is the most favorable one for fishing. The strongest current is *ānthapperuveḷḷamvalivu*, which violently stirs up the seabed. It begins in the month of *chithirai* (April–May) and lasts till the month of *vaikāśi* (May–June), and is said to bring shrimps to the surface as their habitat gets disturbed. The diagram below depicts the main currents.

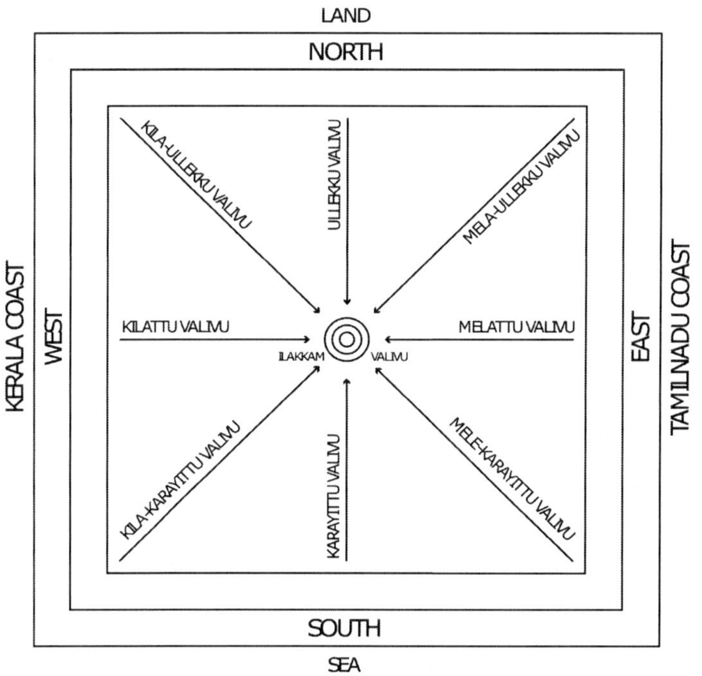

Fig. 4. Currents and their Directions

The celestial phenomenon is the trustworthy companion of a fisherman who is mostly out in the ocean. Beyond about 45 fathoms, the shoreline vanishes from sight, and there is no other landmark like roads or buildings or trees. Instruments like compass have been introduced only very recently. As a result, the fisherfolk have acquired an amazing treasure house of maritime astronomy as contained in the concept of *kaṇiyam*. During the day, the movement of the sun is the sure indication for direction. But cloudy days and the rough season make the sun unreliable. For fishing at night, the position and movement of the moon and the stars play a key role. When signals from the lighthouse also vanish from sight, fishermen keep track of certain stars or planets or constellations, which they call *veḷḷi* (silver) or *meen* (literally it means fish). The frequently referred *veḷḷis* are the following:

Kappalaveḷḷi: It is a cluster of eight stars in the shape of a ship with an anchor. Since this constellation does not change its formation all through the year, it is considered very reliable by the fishermen.

Chōṭṭuveḷḷi: This is a single star that rises in the evening, around the time for night meals. Hence, it gets the name *chōṭṭu* (night meal).

Malaimeen: This is also a single star that rises at midnight from the shoreline and sets in the sea. It is visible around Christmas time, and so is called *Yeśu Nakshatṛam* (Jesus Star). The star is visible in the Malayalam month of *Dhanu* (December–January), usually on the 10th that is likely to fall on Christmas day. To the fisherfolk, this was the star that led the three kings to Bethlehem. It is when the *malaimeen* appears that big fish varieties come out of their deepsea cleft (*pāru)* in search of prey/bait.

Vidiveḷḷi: This single star rises early in the morning, and so is called *vidiveḷḷi* (morning star). It is also called *kaṇṇanveḷḷi.*

Muzhakkāmeen: A cluster of three stars more or less in a straight line appearing like a measuring rod is the *muzhakkameen*. It is visible till *Chittira* 10 (April–May).

Āṛāmmeen (*āṭṭuveḷḷi*): This is a group of six stars in a cluster; it is visible till *Vaikāśi* 10 (May–June).

Kurisuveḷḷi: This constellation shows four stars arranged roughly in the shape of a cross; traditionally astronomers called it the southern Cross. The lower star that forms the stem of the cross is the brightest and points to the South Pole.

Āmaveḷḷi: This is a group of stars appearing in the shape of *āma* (tortoise), and it rises and sets in the ocean close to the southern horizon. It appears in the month of *Āni* (June–July).

Kadaluveḷḷi: This appears a week after the appearance of *āmaveḷḷi*. This too rises and sets in the southern horizon in the ocean. Fishermen predict rough weather when *āmaveḷḷi* and *kadaluveḷḷi* disappear.

A sample of traditional marine astronomy can be seen in the following lines, as narrated by a fisherman.

> If you follow *kuriśuveḷḷi* you can reach *chandālakeṭṭu* (name of a cleft underwater); if you follow *muḷakkāmeen* you will reach *chendadichānkeṭṭu*; if you want to go to the deep sea after nightfall, follow *chōṭṭuveḷḷi*; to reach the shore safe at night, follow *kappalāveḷḷi*; to reach ashore early morning, follow *vidiveḷḷi*.[7]

For ages, humans have relied on stars and planets to guide them across the desert or over the ocean. The regularity and meticulous movements of the heavenly bodies helped them in measuring time or charting directions. Agricultural people studied the seasons to know when to plant or to harvest. Similarly, fishermen used their knowledge of the celestial bodies to determine the proper time for fishing and to keep track of directions in the ocean. The very nature of their occupation on the surface of the wide ocean makes it necessary for fishermen to be good marine astronomers. Astronomers cite more than eighty constellations. The names of these are always culture bound based on resemblance to an animal or an object familiar to that culture. It is natural that the fisherfolk discover a ship or a turtle or a cross in the constellations that guide them.

The Power of Lunar Influence

The moon and the lunar phenomena have a still greater influence in the life-rhythm of fishermen. Their day, in fact, is not confined to exactly 24 hours, nor is the distinction between day and night very marked. The time for going on fishing expedition is determined mainly by the rising and setting of the moon, not the sun. Other factors like *nilavu* (moonlight), *kavaru* (glow), and tidal phenomenon also have their roles. The daily occurrence of tides changes the contours and behavior of the ocean along the coast, thus shifting the time for fishing every day. The gravitational force of the moon, the power behind the tides, results in two *flows* (upward tide) and two *ebbs* (downward tide) every day. In scientific terms, the interval between one flow and the next flow at a particular

---

7 Personal conversation with Johnson in October, 1996.

place is 12 hours and 25 minutes. That means today's flow will be 50 minutes later than yesterday's.

It was Sir Isaac Newton, an English physical scientist and mathematician (1642–1727), who first explained the tidal phenomenon scientifically in the seventeenth century.[8] By his work on gravitation and planetary motions, he could explain how one and the same force of universal gravitation causes planets to revolve around the sun in elliptical paths and acts between any planet and its satellites. This same force acted also between the ocean, on the one hand, and the sun/moon, on the other, thus causing the tidal phenomenon. He also computed the relationship between the distance and the mass of the body. (These were published in his *Mathematical Principles of Natural Philosophy* in 1687). The impact of the moon on the sea is not uniform always. Two factors affect the change: the waxing and the waning of the moon, and the changing distance from the earth to the moon due to the helical orbit of the moon. Thus, there is a double change daily occurring in the ocean due to the moon. Relatively, the sun has only a limited influence over the ocean; the lunar tidal force is 2.2 times stronger than the solar tidal force.

Whether it is the full moon or new moon is of vital importance to the fisherfolk. "It is bright moon today, fish won't come out," they would say. The full moon may bring joy to poets and artists, but to fishermen, it will be a dry day. The net or the line becomes visible to the fish in bright moonlight, and by instinct, they stay away. The fishermen are forced to readjust their work schedule and go for fishing either before the moonrise or after it sets. The launching time shifts gradually every day, either forward or backward according to the changing moon, with fishing reaching its high point before the new moon. What is crucial is the launching time, never the returning time, fishermen point out. Astronomers discuss different systems of measuring time, viz., solar-based, lunar-based, and stellar-based. More significant for our discussion here are the solar and lunar systems of measurement.

The interval between the two passages of the sun across the meridian is a solar day. The mean solar day has 24 hours 3 minutes and 56.55 seconds. As against this, a lunar day is 50 minutes longer than a solar day. The solar year has 365 days 5 hours 48 minutes and 46 seconds, while a lunar year is 11 days shorter than the solar year. The moon takes 29 days 12 hours 44 minutes and 2.8 seconds to complete one revolution around the earth; this is called a *synodic month*. This lunar month is shorter than

8 See *Encyclopedia Britannica*, 15th ed., s.v. "Sir Isaac Newton".

the solar month of 30/31 days, the latter being an arbitrary division of the solar year into 12 parts.[9]

Against this background of a widely accepted calendar that is basically solar, the fisherfolk find themselves at variance. Their daily rhythm of life is centered on the moon and the tides, as discussed earlier, and they follow the lunar calendar both in their occupation and in their daily life. This means that a fisherman's day is 50 minutes longer than a civil day and that he has 11 days less in a year. Moreover, his calendar is marked by daily fluctuation, as against the relative stability of the solar calendar. A fisherman's (lunar) "regularity" is treated as "irregularity" by the mainstream (solar) society. This discrepancy characterizes the fisherfolk vis-à-vis the mainstream society. The whole family shares in this predicament. The womenfolk, though not engaging in actual fishing, cannot escape this dissonance. The work of fish vending women can begin only after the fishing boats land ashore. If the catch is poor, the women have to wait endlessly for more arrivals. Such unpredictability disturbs their household chores too. The long absence of both the parents from home adversely affects family life, care of children, and their education.

To be a fisherman is to be attuned to the rhythm of the ocean that is ever in turmoil. To live, for him, is to live amidst change every moment. The wind and the waves, the moon and the tides, the fish, and the stars—all bring to him every moment the reality of change and of movement. His today is quite unlike his yesterday, and he also knows from experience, that tomorrow will again be different. It is this eternal newness that makes every day fascinating and makes life dynamic and challenging. Yet there is order in this change, he knows; he knows when the next tide would occur, and when the moonlight will vanish. But unfortunately, his "order" is taken for chaos by the mainstream culture that depends so mechanically on an order determined solely by the solar calendar and the

9 For more details, see *Encyclopedia Britannica*, s.v. "Calendar". See also Edward S. Sachau, ed., *Alberuni's India* (Delhi: S. Chand and Co., 1964), chapter 36. The discrepancy between the lunar and the solar calendars ever remained a challenge for astronomers. A lunar calendar is believed to have been in existence in most ancient cultures including the Indus Valley civilization. Agrarian cultures generally relied on a solar calendar due to the importance of seasons for their occupations. However, many cultures adopted a "luni-solar" calendar in which months are lunar, while days and years are solar. The old calendar of the Greeks and the Vedic calendar of the Hindus are examples. The Julian calendar, introduced in Rome in BCE 46, was strictly a solar calendar. This was corrected in 1582, under Pope Gregory, as the Gregorian calendar, which in course of time came to be accepted universally as the system of measuring time. Some ancient peoples such as Jews, Muslims, and Hindus still make use of the lunar calendar but as restricted to the realm of religion and rituals.

wristwatch. A calendar that does justice to the complexities of the coastal time and space is nonexistent today. This discrepancy has far-reaching implications. The high rate of school dropouts, failure in timely repayment of bank loans, and the difficulties the young jobseekers face in adjusting to a different work schedule and work culture are some examples.

Could it be that the fisherfolk are caught up in a conflict of two forces, being pulled from different directions by the lunar and the solar phenomena? Does their life amidst two contrasting systems of time consciousness make them victims of two opposing worldviews? How do the fisherfolk manage to deal with this conflicting situation to make sense of their life? What role does their Neithal Catholicism play in sustaining them amidst these conflicts and discrepancies? Questions like these emerge in the process of our inquiry into coastal life and religion.

Oriental people generally were followers of a lunar calendar. The Kollam era was in existence in the Kerala society for ages. Under the strong impact of a colonial past and of a process of westernization, many societies adopted the solar calendar (Julian calendar) for administrative purposes. Thus, we find the coexistence of both the Julian calendar and the Kollam era in the mainstream Kerala society with hardly any conflict. But it was very different in the case of the fisherfolk who have been experiencing the conflict deeply in their daily life. The dominant culture and its time consciousness are heavily weighed against the fisherfolk. The prolonged influence of the Western powers and of Iberian Catholicism provided them with little help in this regard.

The Wisdom of *Kaṇiyam* and the Ockhi Disaster

Moving among the coastal villages of Thiruvananthapuram soon after the Cyclone Ockhi, I couldn't help asking fishermen three vital questions. One, did the inherited wisdom of *Kaṇiyam* fail the fishermen in detecting the Cyclone and its destructive power? Two, did the deadly impact of Ockhi change their attitude to Mother Ocean and to fishing itself? Three, do they feel that their faith in God and their religion failed to come to their assistance in the midst of such a calamity? I could discuss these with some master fishermen, with some fishermen who had survived Ockhi, and also take these up in focus group discussions.

The elders said that in their lifetime they never had experienced a cyclone of this magnitude, and so it was not easy to interpret the abnormal behavior of the ocean. The wind from the landmass was misleading, and so also the strong pull from under water. Many of them had noticed unfamiliar signs on their GPS, and sensing danger, returned to shore safely. The elders in some boats advised the crew to stop work and rush back;

but the thoughtless bravery of the youngsters prevailed, only to end up in danger. They all testify that the problem was not with *kaṇiyam* but with the men who put it to use. The new generation of younger fishermen either ignore or fail to read the signs of the ocean when unusual changes occur. "Why should we blame *kaṇiyam* for this human foolhardiness?" the survivors ask. One of them quoted the proverb: "He who remained watchful survived; he who slept got lost." *Kaṇiyam* presupposes deep reverence for Mother Ocean and sensitivity to the pulse of her movements and changes. They are pained that this reverence is vanishing. The new generation is inclined to rely more on technology and gadgets like GPS and wireless communication, but to their surprise, wireless sets were switched off on many boats. "*Kaṇiyam* is not at fault, but we are," is their conclusion.

Is there a discernible change in the attitude of fishermen to Mother Ocean after the Ockhi disaster? Did Cyclone Ockhi shatter their faith in the goodness and bounty of Mother Ocean? The survivors narrated the trauma they have gone through after escaping the near-death experience. They are haunted by the sea with its mighty waves rushing to swallow them and tossing them around mercilessly. Many of them are still unable to sleep. They dread the very sound of the wind and the rain that fills them with fear, and it holds them back from venturing into the sea again. Fear and reluctance to go for fishing is a common symptom in survivors. They dread both nightfishing and deep-sea fishing and engage mostly in shoreseine operations. They were always aware of the ambivalence of the Ocean with her benign and fearsome features, but now after Ockhi, the fearsome nature seems to dominate. I found many survivors still struggling to come to terms with this confusing mindset.

At the same time, the fishermen, both survivors and others, are emphatic in their unshaken faith in Mother Ocean and her goodness. "Kadalamma will never endanger us," they asserted in a focus group discussion; "disaster is due to our irresponsibility, we are to be blamed for it. It is our faulty lifestyle that is at the center of such disasters. Many younger men are addicted to drinking and to drug abuse, and they lose their ability to discern the 'signs' (*kaṇiyam*) properly." The sense of reverence for the sea that fishermen always held sacred is on the decline. Elderly fishermen used to drink, but never before going fishing. So, the new generation is losing the inherited knowledge of *kaṇiyam*, and this needs critical study. The majority of those who lost their lives were addicted to drinking and had a big belly, they point out. People refer to certain discernible changes in the lifestyle and attitude of fishermen after Cyclone Ockhi. One of the survivors was very articulate when he said, "Many people stopped, or at least reduced, drinking and drug abuse. There is a growing realization that our lifestyle needs correction. Our faith in God and in religious practices has only increased, not diminished. We see our saving experience as coming from God, whether through a ship or a helicopter."

An undying hope marks the fishermen of the Thiruvananthapuram coast. The words of Raymond, a fisherman from Poonthura, illustrate the general sentiment of many survivors: "I gave up all hope, and was prepared for death. My only prayer all through was, 'Take me ashore, Lord'. The sight of the ship brought us hope again, and I was convinced that the ship was sent by God." Alban, another survivor, joins in: "Eight of us were hanging on to a piece of rope on the capsized boat for two full days. We tried in vain to catch the attention of a passing ship to notice us. Then a helicopter came in our direction but passed us by. Losing all hope and ready to die, I prayed from my heart: 'Lord, you take care of my three little daughters'. But to our surprise, the helicopter returned and began to pull us up one by one." Lawrence is convinced that he was given a second life since he was counted as "dead" by his family and village. Special prayers were going on at home on that fateful day, and he believes that his "rebirth" was the result of their prayers.

CHAPTER 5

LEARNING FROM THE RITUALS OF COASTAL LIFE

The notion of ritual has been central to the inquiry in the fields of religion and culture for many decades. This is because rituals, as symbolic actions, make observable the worldview of a people, including their view of the realm of the sacred. Victor Turner uses the term ritual to refer to "prescribed formal behaviour for occasions not given over to technological routine, having reference to beliefs in mystical beings or power."[1] The insight of Carl Jung on the role of the collective unconscious in the formation of rituals has helped to validate the "ritual route" to study religion.[2] Having a collective performance dimension, rituals are accepted as useful keys to grasping the cultural/religious contours of a community.

Generally, a religious ritual is regarded as an action that is performed and is distinguished from the conceptual aspects of religion such as beliefs, creeds, and doctrines. Durkheim tried to analyze religion as both beliefs and rites, thus giving analytic primacy to beliefs. However, a clear distinction between thought and action does not resonate with the experience of people not having a speculative tradition. For them, rituals constitute the medium through which both conceptual and behavioral components get merged. Ritual, as Catherine Bell puts it, "is a type of critical juncture wherein some pair of opposing social or cultural forces comes together. Examples include the ritual integration of belief and behaviour, tradition and change, order and chaos, the individual and the group, subjectivity and objectivity, nature and culture, the real and the imaginative ideal."[3] Thus, rituals serve as a window on the inner dynamics by which people make and remake their sacred world.

1 Victor Turner, *The Forest of Symbols: Aspects of Ndembu Ritual* (London: Cornell University Press, 1977), 19.

2 See Joseph Campbell, ed., *The Portable Jung* (Penguin Books, 1976), chapter 4.

3 Catherine Bell, *Ritual Theory, Ritual Practice* (New York: Oxford University Press, 1992), 16.

The term ritual is widely used in disciplines such as cultural anthropology, sociology, or theology, but is not always uniformly understood. Some use it in a broad sense to refer to all "culturally defined sets of behaviour" irrespective of their explicit religious or social content. Sociology of everyday life tries to understand rituals in a system of daily human interaction, thus assigning the term to all symbolical dimension of human behavior.[4] Traditional theologians, on the other hand, tend to link rituals narrowly with liturgical ceremonies and a religious cult. This chapter prefers to take an intermediate position. First, the cultural and religious horizons seem to interpenetrate in the realm of rituals, thus making it difficult to stamp certain rituals as purely religious and others as nonreligious. Second, rituals are viewed not only as a means of expressing collective beliefs and ideals but also as generative of beliefs and ideals simultaneously. The *rites of passage*[5] in the life of coastal people do not become the focal point of this chapter as these are largely subsumed into the sacramental framework of Catholicism. So, we pay attention to the broad categories of occupational rituals, domestic rituals, and curative rituals in the coastal world. The method involves three main steps: (a) consideration of the observed features of the rituals in the context of everyday life; (b) eliciting the explanations or interpretations of these rituals by wisdom figures in the community; and (c) exploring broader associations with scholarly viewpoints and drawing inferences.

Fishing as Ritual Performance

Anthropologists have classified traditional human societies into different categories, mainly based on methods of securing food. Raymond Firth distinguishes four major societal types: (a) food gatherers, hunters, and fishers, (b) pastoralists, (c) agriculturalists, and (d) artisans.[6] Fishing is fundamentally a hunting and gathering operation. Quite contrary to an agrarian context, the psyche of a fishing community would display a gambling instinct. Fishing in the traditional way is a short-term operation that is usually complete in a day or two. There are days when the catch is plentiful, but also days when the net is empty. This would explain why the fisherfolk are staunch believers in fate or providence often giving expression to practices that may appear superstitious. In fact, they draw

4 See Erving Goffman, *The Presentation of Self in Everyday Life* (New York: Doubleday, 1959); Roy Turner, ed., *Ethnomethodology* (Middlesex: Penguin, 1974).

5 Arnold Van Gennep, *The Rites of Passage*, trans. M.B. Vizedom and G.L. Caffee (Chicago: The University of Chicago Press, 1960).

6 Raymond Firth, *Human Types: An Introduction to Social Anthropology* (New York: Mentor books, 1958), 48.

strength from these beliefs and practices in order to survive amidst risk and danger. Fishing for them is much more than merely catching fish from the sea, whether for their own consumption or for the market. It is the most central human activity that expresses their basic beliefs and ethos, their very identity and life orientation. The series of ritual acts meticulously performed in a fishing expedition has significance not only for their work but also for their entire life. Fishing thus becomes a sacred activity. In and through this, the fisherfolk make and remake their world that is simultaneously secular and religious. So, it is appropriate to call fishing a ritual performance.

From a ritual point of view, traditional fishermen generally engage in either of the two modes of fishing: operation in the shallow waters close to the shore, and operation in the deep waters. Fishing using shoreseine nets (*kampavala*) is a typical example of the former, while gill nets and trawl nets, besides hook and line fishing, are employed in the second category. Technological innovations such as lighter boats made of plywood or fibreglass and the use of outboard engines have transformed fishing in recent times, although traditional wooden canoes and the age-old catamarans are still widely used. Rituals that are specific to the operation can be seen in both the modes. Shoreseine fishing is a collective operation that requires a large number of fishers, often twenty to forty, working in two teams pulling the net ashore from the shallow waters. People of all age groups can participate in it. What makes it attractive is the collective singing of folk songs to keep rhythm for pulling in unison. This mode of fishing operation has remained practically unchanged for generations enabled by the ritualistic aura it has acquired over generations of fishers. The sudden slanting of the ocean floor along the Thiruvananthapuram belt has made this mode of fishing very popular in many fishing villages.

Eilamidal is a ritual ingredient of shoreseine fishing operation. It involves the rhythmic singing of folk songs by the crew members while pulling the shoreseine net ashore. Short versions of the same can be heard while pushing the boat into the seas and while pulling it ashore. A sample of *eilamidal* as it is sung in Vizhinjam is given below.

(Begins in slow rhythm)

eilavalē, lovalamma, eilavalē, lovalamma
lovalamma, eilavalē ...

chorus: eilavalē

eleilayya, elavalayya, ōvalayya, oveilayya
eleilayya, elavalayya ...

chorus: eilavalē

(The rhythm reaches a crescendo as the catch nears the shore)

ōvale, ōvale, ōvale ...

ōvale, ōvale, ōvale ...

(*in unison, no chorus*)

Eilamidal is a rich source material in understanding the religious world of the fisherfolk. In the relative absence of historical documents or written literature, this form of oral literature makes available an unbroken tradition of the neithal past handed down through generations in relative purity. Two reasons support this view: the shoreseine is a very traditional mode of fishing that has undergone little change over time; change is slow to come in this mode of fishing due to the employment of a large number of people, led by the elders. Second, the rhymes used as occupational songs show little variation between fishing villages.[7]

"While engaged in fishing, a fisherman sees nothing but the float of the net," (*Mukkuvanu meen pattāl muthappayil kaṇṇu*) is a popular saying. Persons and relationships all vanish from his thoughts while the catch absorbs him fully. This is evident in the ritual evaluation (*cēluparachil*) that takes place at the end of every fishing operation. After pulling the catch ashore, the crew members stand around and review the operation in detail—the manner of work, the time chosen, the depth and direction, the coordination, etc. It is a sort of ritual evaluation and may take an aggressive turn if the venture was frustrating. The blame naturally falls on someone or another. To an outsider, it may appear like a violent group clash where aggressive and filthy language and gesture are liberally used. In ten or fifteen minutes the "exercise" is over, and the gang members get back to their routine tasks—sorting the fish for auction, cleaning the net, and mending it. The effect of *cēluparachil* is not allowed to carry over and spoil their social relationship; what happens in the sea is not to be carried ashore, they had been taught. However, *cēluparachil* serves a tremendous social-psychological purpose by acting as a ritual forum of catharsis, especially after a failed fishing venture.

Fishermen routinely take their turban off from their head and tie it around their waist while launching their boats. "Kadalamma (Mother Ocean) does not like that," they say, and "she will push the boat back to the shore in fury if we fail to do that," they believe. Removing one's turban is part of paying homage to someone in the Kerala tradition. One is not supposed to wear footwear or carry cash while going fishing; Kadalamma will take him for a rich man if she finds him with money, and won't be

7 Some Christian fishing villages in the Kanyakumari district of Tamil Nadu have modified the rhymes to give them a Christian flavor.

generous with him. (This is not strictly followed in the changing work context of today). The boatmen pause for a while as they reach the old church to bow and pay homage to their Amma (Cintāthira Māta), the patroness of the church. Rituals are prescribed at the launch of a new boat or while fitting an outboard engine or a new gear. Blessings from a priest are sought, and coconut is broken (a Hindu custom) and distributed with jaggery. Toppling a water pot (*niṛakudam maṛikkal)* is another related ritual. A pot that is filled with water is kept at the front door, and the person carrying the new item must topple it with his foot and proceed without looking back.[8] In the south Indian tradition, a pot that is full and overflowing is a sign of prosperity and plenty, and to the fisherfolk, it is an auspicious act for a bountiful harvest.

Taboos and Omens

Belief in taboos and omens is strongly held by the fisherfolk, and various rituals make it tangible in their everyday life and work. The relationship between Kadalamma and the womenfolk seems to be ambivalent. The ocean is always feminine; she is Kadalamma, Mother Ocean. But fishing has ever remained a male centered activity. The seashore is the occupational world of the men folk, and so women are conspicuously absent. The exception is the presence of elderly fish-vending women who come to buy fish at the landing center. Women avoid the beach especially when fishing is in progress. It is taboo for women to enter a fishing boat and go to sea, except on a Christmas day. Kadalamma may get angry and the boat may capsize if women enter, they believe, and they narrate instances of such occurrence in the past.[9] Women should not stand on the beach with their hair dishevelled; "Kadalamma does not like that."It is a bad omen for a fisherman to see a woman with dishevelled hair on his way for fishing. The sight of a widow in one's way is also considered a bad omen. Coastal women too are very conscientious about the existing taboos and prohibitions in approaching the sea. Interviews with women informants revealed how powerfully they have internalized these prescriptive norms, particularly norms of purity and pollution. If reverence marks the mindset of the menfolk toward Mother Ocean, it is the spirit of purity and pollution that keeps the womenfolk away from her.

Fishermen have been taught that they should avoid the sight of an empty pot on their way to fishing. The pathways are frequented by

---

8 The same ritual is used in marriage ceremonies where the pot is toppled by the bridegroom when the new couple enters the house.

9 Such taboos exist also in fishing communities elsewhere, as ethnographic records reveal. See Firth, *Human Types*, 48. There are exceptions too.

women who carry empty pots to fetch water from public wells or water taps. It is not uncommon to see such women rushing to hide their pots if they happen to come across a man on his way to fishing. On the other hand, if it is a pot filled with water, it is a good omen. On his way to fishing, a fisherman is not supposed to look back, nor should he be called from behind. Decorum must be maintained at home when the menfolk are away fishing in the sea, and the women and children back home avoid shouting or quarrel. If children happen to make noise or cry, the mother would rebuke and remind them that their dad is out in the sea.

Belief in *śāpam* (curse or spell) is prevalent in the community, and cats and dogs are particularly significant. Cats are a common sight wherever fish is handled. If a cat is allowed to eat raw fish kept at home, it may result in "cat-curse" (*pūchaśāpam*). This is why even the waste pieces of raw fish are not thrown to cats. If their nets do not fetch enough fish for many days at a stretch, fishermen conclude that the reason is "cat-curse," and so will resort to remedial mantric rituals. If a fisherman sees a cat crossing his path or coming in the opposite direction while going fishing, it is a bad omen. If it is a black cat it is worse, and he may avoid fishing altogether that day. A popular saying is that it is a bad omen if a cat crosses your path, while it is a good omen if it is a dog crossing your path *(pūchavēṭṭa puḷikkavēṭṭa; nāyvēṭṭa nallavēṭṭa)*. Mixing a paste of "*vayampu*" (*Acorus Calamus*) with Holy Water and pouring it over the gear is a ritual used for "cat-curse."[10] Belief in "evil eye" *(kampēṛu* or *kaṇṇupedal)* and black magic (*gūdopadravam)* is also widespread. If the catch is surprisingly poor, a fisherman would say that someone has done black magic (*mantravādam)* against him. In times of illness or misfortune, fishermen easily ascribe these to the spirits of the dead *(peiy)* that they believe would haunt and disturb living persons. Most fishermen have the habit of chewing betel leaves and spitting red liberally all along the way. To them it is not a bad habit; rather, it is a necessary ritual that protects them from evil spirits besides keeping them awake and alert for work at odd hours of the night. They are convinced that if evil spirits come to harm them, a mouthful of betel juice in full force will keep them away.

How do the fisherfolk deal with "evil eye"? The ritual they employ is called *kampēṛunōkkal* (checking for an evil eye). It is usually done in two ways. In the first way, some table salt, some mustard seed, the outer peel of onion, and some green chili (3 or 9 in number) are bundled up in a paper and moved around the affected person in a twirling motion; then,

[10] In some coastal villages another ritual is performed, especially when favorable winds fail. One person would tie the root of a particular tree on to his leg, and run at night, naked, in the direction in which wind is desired. He would recite the Christian creedal formula in the reverse while running.

after spitting on it three times, it is thrown into the glowing embers of charcoal. If it bursts and produces a foul smell, it is the proof for evil eye. In the second way, some special salt is taken and moved around the affected person in a twirling motion as in the first way and is placed on the glowing embers of charcoal made from three coconut shells having holes. The salt melts in the heat and spreads. People watch the abstract shape of the spread and judge if there has been evil eye. Then the spread is taken to a crossroad and is left in the middle of the road to be trampled upon. Another important ritual is *erikkilaadikkal* or beating the net with twigs of *erikku* plant (*calotropis gigantea*). When a net fails to fetch fish for many days at a stretch, the owner concludes that some evil has befallen, and so resorts to this ritual. The fishing gear is kept in a closed room in the dark and is beaten fiercely by a young boy or boys in total nakedness, shouting *pōpiśāśē* (go away, you evil spirit), and using all kinds of abusive words.[11]

Death and Life Beyond

The life of fishermen is characterized by an ever-impending danger of death that is closely linked with their occupation of fishing. The rough monsoon season is accident prone often leading to loss of life in the sea. Natural disasters like cyclone and tsunami are not uncommon and may result in the death of hundreds of fishermen. Many are declared lost in the sea as their bodies are never recovered. The death of a fisherman is a tragedy not only for the person and his family but equally for the whole coastal village. Such instances bring a pale of gloom over the village with the family and their relatives and neighbors gathering to keep vigil in prayer for days for the safe return of the missing person even when no hope is in sight. This explains why rituals related to the dead assume great importance for fisher communities.

After the death of a person, the close relatives stay in mourning for a period, usually seven days, and this ritual is called *chiṛakākkal.* They stay away from work and other duties during this period. If the dead person is a married man, the wife has to go through a series of rituals. As the dead body is being prepared for burial, the wife is taken away by her in-laws and her head is covered with a white cloth; this ritual is called *muṭṭankiyidal.* On the seventh day, the *tāli* (wedding necklace, the symbol of married state) of the wife is broken and removed by her brother-in-law. This

---

11 *Erikku* has an important place in many mantric rituals in the Hindu tradition; its stem, roots, or flowers are prescribed in many rituals, especially to propitiate Kāḷī, Bhagavathy, or Gaṇapathy. All the 27 stars (*nāḷ*) have a corresponding tree (*nakṣatravrikṣa*) in the mantric tradition; *erikku* is the tree for *tiruvōṇam*.

ritual is called *thaliyural*. The sister in law then applies oil on the woman's head and takes her for a ritual bath and dresses her in new clothes; this is called *muṛanadathal*. For a woman who lost her husband, it is painful to go through these rituals. These rituals are on the decline today as the younger generation looks at these customs as evils to be done away with. An account by Thurston a century ago regarding the funeral rites of the Hindu fisherfolk would throw more light on these rituals. He writes:

> The dead are as a rule buried.... The corpse is bathed in fresh water, decorated, and placed on a bier. The widow then approaches, and with a cloth over her head, cuts her *tali* off, and places it by the side of the corpse... . Death pollution is observed for seven days, during which the son abstains from salt and tamarind.... On the eighth day, or sometimes the fourteenth day, the final death ceremony is performed.... The eldest son is expected to abstain from shaving his head for six months or a year.[12]

It is obvious that the traditional customs and rituals continue to be upheld in the lived religion of the Catholic fisherfolk. Many Christian rituals, on the other hand, have got assimilated into the traditional neithal ritual configuration of the community. A Mass in honor of the dead ancestors is a requirement when an auspicious event takes place in the family, viz. marriage, moving into a new house, or joining a new job. The idea of invoking the spirits of the ancestors and seeking their blessings thus indicates strong residues of ancestor worship from a tribal past. When a fisherman is reported missing in the ocean, the people say he has been taken away by a *Kadal Muni* (Sage of the Ocean). One informant testified: "*Kadal Muni* takes them to some far away mountain and keeps-them there. He would teach them many skills like the art of medicine and healing. Sometimes they come back to the village disguised as wandering medicine men *(Lātaguru)*. I myself had an encounter with one of them." That those who are dead continue to exist, either as spirits *(peiy)* or as wandering medicine men or any other, is a deeply held belief. It points to their faith in a mode of existence after death and a hope that death is not the final end of life. This hope may be the driving force behind the relentless struggle of the fishing community before the danger of death, and the secret behind their resilience after a disaster.

The All Souls Day that the Catholic tradition celebrates on November 2nd every year has special significance in coastal communities. The ritual of blessing the graves *(kuzhiōthal)* involves elaborate preparations in the cemetery lasting many days. Wet sand from the seashore or red mud from elsewhere is brought to give a facelift to the graves/tombs of the departed family members in the parish cemetery. Each grave is

12 *Castes and Tribes of Southern India*, Vol. V, 116.

decorated with flowers and green leaves and lit up with candles. Relatives squat around the grave in prayer patiently waiting for their turn for the priest to arrive. It is not uncommon to see the same grave being blessed over and over again depending on the number of coupons the family members have purchased. The souls of those who die untimely, like the death in the sea or an accident, are believed to wander on this earth as *peiy* (spiritual beings) and may disturb the living relatives as *bādha* (possession). Coastal beliefs relate the fate of the dead with the concept of *peiy* and *muni.*[13] This is why prayers and rituals are duly offered for their speedy deliverance. This belief partly explains why fishermen worry about recovering the body of those who die in the sea, as has happened during the Cyclone Ockhi, and giving them proper ritual burials. Hindu fishermen also offer *pujas* and rituals for days after death and build small tombs near the house "to settle the ancestors." In the Catholic tradition, it is transferred to the common cemetery with little change in the basic belief system.

The vulnerable life context prompts fisherfolk to take refuge in various forms of holy offering *(nercha)* as they meet with misfortune or problem situations. Salt and pepper *nercha* at the statue of Cintāthira Māta is a common sight at the parish festival at Vizhinjam. Other forms of the offering include the offering of a token boat made of silver, dressing up of the statue of Mary with a new veil, and distribution of rice porridge to devotees. The Catholic fisherfolk make offerings not only at the Christian church but also at the nearby Hindu temples and Muslim shrines with equal devotion. They promise ritual gunshots (*vedinercha)* at the nearby Amman temple to recover lost articles.[14] A *nercha* is offered at the Thaṭṭātheru temple to punish someone or to take revenge on someone. An offering of "oil and wig" (*eṇṇayum thiriyum*) is made at the same temple in order to receive favors. Fish vending women and cyclists, on their way to the inland markets, stop at the temple to make offerings for good business, and on their return too they make offerings

13 See Louis Dumont, *A South Indian Sub-Caste: Social Organization and Religion of the Pramalai Kallar* (Delhi: Oxford University Press, 1986), 449–450, for similar notions. While studying the *Kaḷḷar* caste of Tamil Nadu, Dumont observes two categories of spirits, gods, and demons. Gods are spirits that receive a regular cult and are usually confined to a temple or cult place. Demons are believed to be wandering and not confined to a temple or cult place. Demons can be either *peiy* or *muni. Peiy* has the appearance and stature of human beings but are rarely seen; *muni* (literally, saint or ascetic) are gigantic figures with a terrifying look, huge teeth, and braided hair. The latter haunt trees or paths and are more powerful than *peiy*. The marine fisherfolk share many aspects of the same belief system.

14 The *vedinercha* twice a week at the temple was discontinued after the communal clashes at Vizhinjam in 1995.

in thanksgiving. According to the temple committee, a major portion of the collection at the temple comes from the Catholic fisherfolk. The Muslim shrines also get a fair share of offerings from the Catholic fisherfolk. The two tombs of Muslim saints in an ancient shrine on the beach near Vizhinjam are revered by Muslims and Christians alike, and the miraculous well nearby is a sacred spot for all. One of the sages buried here is invoked to take vengeance on one's enemy. The offering of "bucket and rope" *(pāḷayum kayarum)* at the well is said to be effective in curing children's illness.

Mantravādam—Its Significance in Coastal Rituals

Protection against evil is a major preoccupation of people in fishing communities. They believe that misfortune is caused largely by the evil spirits active all around them, and so resort to magical rituals, popularly known as *mantravādam*. Since it is usually done in secrecy, people rarely admit to its use. This makes any inquiry into the realm of magic rituals quite risky and difficult. Many mantric rituals are remedial measures to undo the effect of evil or to ward off evil and to ensure good luck. Interviews with a wide spectrum of people made it clear that the rationale for most of these ritual practices lies in oblivion. Some key rituals and their significance will be discussed in this section.

If one boat alone gets a bumper catch while others return empty, people normally attribute the catch to *mantravādam*. If a fishing operation ends up in failure with a poor catch, then they say fish has been carried away by evil spirits. Practitioners of mantric rituals are able to propitiate Kadal Vairavan (a god of the ocean) and to stir up the sea to prevent a good catch, people believe. Vairavan is the protector or supervisor of fish wealth in the ocean according to Hindu fishermen. The word Vairavan (*Bhairavan)* means frightful or terrible. It is the name of one of the eight terrible (*ugra* or *ghora*) forms of Rudra-Siva, especially worshipped by outcaste groups.[15] Fishermen secretly tie amulets (*eilass*) to the boat or the gear.[16] In anthropological usage, a charm that has a protective function is "amulet," while a charm that carries good fortune is "talisman." Belief in the efficacy of *mantras* is a central part of the ritual world of the fisherfolk as its widespread practice, often in secrecy, testifies.

Mantravādam is usually associated with the Hindu religion. But some Catholic fishermen practice it too either in a uniquely Christian way or

15 See Margaret and James Stutley, *A Dictionary of Hinduism: Its Mythology, Folklore and Development 1500 BC–AD 1500* (Bombay: Allied Publishing, 1977).

16 *Eilass* (amulet) normally consists of *mantras* written on a copper foil and enclosed in a tiny metal casket.

as per the Hindu system. Mariadasan, 30, a Catholic fisherman practitioner, has been using both the systems. "To mix up both traditions is dangerous", he says; "one has to stick to any one tradition. I am planning to confine myself to the Christian way." The difference between the two traditions is an interesting area for study. The Hindu tradition is said to be based on the Hindu scriptures and invokes Hindu divinities like Śiva, Vairavan, Kāḷī, Durgā, Mohinī, Bhagavatī and Mādan. It is also more expensive since elaborate rituals and complex settings are prescribed. "The Hindu gods need to be fed sumptuously, and some may require even blood," says Mariadasan. Besides, it requires ascetical practices (*tapamirippu)* and recitation of mantric verses usually for 108 or 1008 times or 5008 times for a specified period. Vairavan (Bhairavan) who controls fish wealth is invoked to ensure a bumper catch. The name of Śiva is mentioned in almost all *mantras* irrespective of the deity it addresses. The general belief is that Hindu mantric rituals are more effective if its purpose is to do harm or take vengeance. The requirements for Christian mantric rituals, on the other hand, are limited to the recitation of the *mantras* with meditation. Commonly invoked are Christian saints like St. Antony, St. Michael, St. Sebastian, or at times the cross itself. The general belief is that Christian mantric rituals can surpass the power of those in the Hindu tradition.

"Where did the Christian practitioners learn the art?" I inquired. Davidson, 45, said that he went to the forest and learned it from a forest dweller. Mariadasan learned the first lessons from his elder brother, and later from a Hindu guru. Varghese spent a year under a guru in Tamil Nadu learning the art. The learning has been mostly from Hindu gurus. All believe firmly that this art cannot be learned without a guru (*guruvillā vidya kuruṭṭu vidya).* The mantric verses are usually kept in handwritten collections or are handed down through oral tradition. The common belief is that what is earned through *mantras* will be lost sevenfold (*mantram kondu kittiyathu ēzhumadangu pōkum).* Therefore, fishermen spend the earnings thus received very quickly, and avoid keeping it as liquid cash. Fishermen resort to mantric rituals for three reasons: to bring good luck, whether in fishing or elsewhere; to remove or to ward off evil, especially caused by gods of other religions; and to harm one's enemy. Many fishermen admitted that they were victims of the third, though nobody would admit practising it that way. The Catholic tradition officially disapproves practice of *mantravādam* by its faithful. That is why people practice it in utmost secrecy unknown to the official church. "*Mantras*, whether Hindu or Christian, are always sinful. But since I am already counted as a practitioner, and since this has become my profession, I have to continue with it," confessed Mariadasan. He expresses a feeling of guilt in this confession, and it may be the same for other Catholic practitioners and their clients.

We shall now take a closer look at two mantric rituals popular among coastal Catholics; one is the *Plate of St. Antony* and the other, the *Wheel-Seat of St. Michael.* Both these reveal the persistent attempt of the coastal people to synthesize the Indic mantric tradition with their practice of Catholicism in a way that makes meaning for their life.

St. Antony's Plate (Antoṇiār Chit)

Its main purpose is to ask for favors. Emili *vaidyan* who had been practising *mantravādam* for many years narrated to me the process of preparing St. Antony's Plate. It consists of a copper foil of 3 1/2 x 3 1/2 inches in size, with a *mantra* written on it with a stylus. The *mantra* is addressed to St. Antony asking for a particular favor, or, more often, to get a good catch. In either case, it seeks protection from the evil spirits, explains *vaidyan*. The foil with the inscription is placed before the picture of St. Antony and prayed upon for 13 successive days and is handed over to the devotee. Fishermen believe that if the foil thus prayed over is kept in the house, it would keep away all evil spirits, or if it is tied to the net, it would ensure good catch. Usually, the foil is encased in a metallic casket (*eilas*) to be worn by the devotee on the body. Emili *vaidyan* relies on a prayer manual in Tamil with the title "St. Antony's prayer for the destruction of the enemy" (*Antoṇiār Śatṛusamhāramālai)* for appropriate formats. A closer scrutiny reveals that the plate and the *mantras* are modelled after the *yantras* in the Indic tradition.

A close examination of one *mantra* reveals the following:

- The general structure is the same as that of any Hindu *mantra.*
- Invoked in the *mantra* is St. Antony (once) or the Holy Cross (twice), or Lord Jesus (once).
- Śiva is mentioned four times in the same *mantra.*
- The divinities are invoked with the request to lead all fish into the net of the devotee and to protect the net.

Why is Śiva, a Hindu god, placed along with St. Antony and Lord Jesus in the *mantra*?[17] Is it a deliberate act or an oversight? A plausible explanation would be that an existing *mantra* invoking Śiva was remodelled

---

17 See Wendy Doniger O'Flaherty, *The Origins of Evil in Hindu Mythology* (Berkeley: University of California Press, 1980), 277. Śiva is essentially a non-vedic god, but combines in him various traditions; viz. the Indus Valley Paśupati, a tribal god of destruction, a *Vṛatya* ascetic, the Agni of the Brāhmanas, and various other local divinities all grafted on to the Rudra of the ṚgVeda.

as a Christian *mantra* addressed to St. Antony, and some remnants of the original *mantra* were left behind unnoticed. St. Antony is considered the special patron of seagoing fishermen in the Thiruvananthapuram coast. "St. Antony has performed thirteen miracles (*puthuma*) while Christ has only twelve to his credit; the thirteenth one is especially for helping us, the fisherfolk," says Emili *vaidyan*.

Cakram of St. Michael (Mikhēl Mālakāyude Iruppu Cakram)

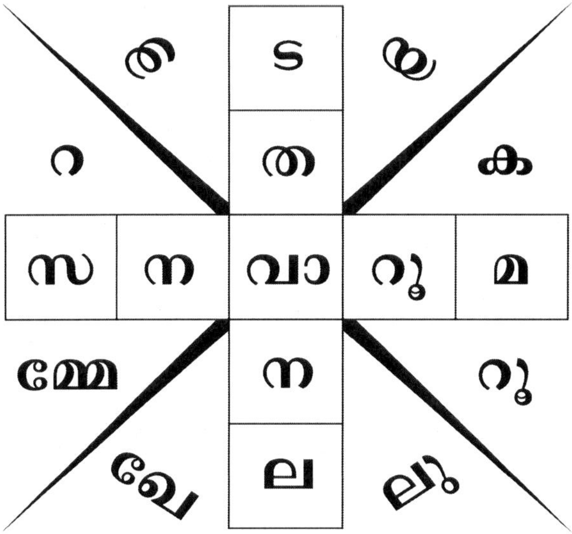

Fig. 5. ***Cakram*** **of St. Michael**

Cakram or wheel-seat of St. Michael, locally known as *Mikhēl Mālakāyude iruppu cakram*, is aimed at the destruction of one's enemies. The wheel-seat consists of a diagram in the shape of a cross with blades between its arms, with 17 alphabets written in Malayalam in a designated pattern. To the untrained eye this may look like a jigsaw puzzle, but to the practitioner, it is a very powerful *mantra*. If properly rearranged, the letters appear thus: ***vāru ma ka yaṭathavā, vāthaṭatharasa a vā, vā a samēkhē la navā, vāna la luru ma ruvā***. The meaning, as given by Panidasan, an active fisherman and a practitioner of *mantravādam*, can be paraphrased as follows:

> O heavenly person, you are the heavenly angel St. Michael
> O holy person, make me also holy
> You who are present in the pillar (wheel?)
> By your holy cross remove from our path
> All the evils of the wicked devil.

The *cakram* is the seat of Archangel Michael built of letters in six angles with a cross and four swords. St. Michael is believed to be sitting at the center of the wheel, and hence the name. The wheel is ever rotating; it is alive. While rotating, it cuts to pieces every enemy. So it is very powerful, affirms Panidasan. This wheel-seat too, in its structure and symbolism, resembles a *yantra* in the Indic pattern.

Śatṛu Samhāra Māla: Protection against evil

Many Catholic houses along the Thiruvananthapuram coast are adorned with a framed diagram of mystical lines and symbols fixed over the outer doorpost. It resembles a *yantra* that is placed in a way similar in Hindu homes or establishments with the difference that the former is usually drawn on paper instead of a copper foil (see figure). People believe that this diagram has the power to keep away all evil spirits. Some people stated that they fixed it after a sudden death or recurrent sickness troubled the family. But hardly anybody could explain the meaning of the diagram or how it operated. The diagram consisted of a cross with jumbled alphabets of Tamil written within, and pictures of three saints—St. Antony, St. James, and St. Michael—and of Virgin Mary drawn around it. A picture of the Sacred Heart of Jesus is placed in the center of the cross

Fig. 6. *Satrusamharamala: The Yantram*

and ten small patterns are drawn around the cross. These patterns resemble Indic *yantras*, but an obvious variation is that the Hindu symbol of trident (*triśūl*) has been transformed into a cross.

I sought the help of a practitioner to decode the jumbled-up alphabets in the diagram. A basic *mantra (mūlamantram)* is repeated many times, and it is an invocation of the Holy Trinity in the mantric pattern. The subsidiary *mantras* written in the columns of the cross are addressed to the three saints mentioned and to Mary as well as to the Heart of Jesus. Four more *mantras* are written at the four corners of the diagram, all addressed to the Holy Cross. Thus, totally there are nine *mantras* around the cross if we ignore those in the small independent patterns. These *mantras* are written in readable language on the reverse of the sheet. Directionality is very important in the construction of this mantric diagram. The first *mantra* in the center refers to the four corners, eight directions, fifteen ends, the sky, the earth, and the valleys, and asks for protection from evil spirits coming from all these directions. This binding of the directions is called *keṭṭu*, and is repeated in the next four *mantras* entrusting the protection of each corner to the three saints and to Virgin Mary—east to St. Antony, south to St. James, west to St. Michael, and north to Virgin Mary.

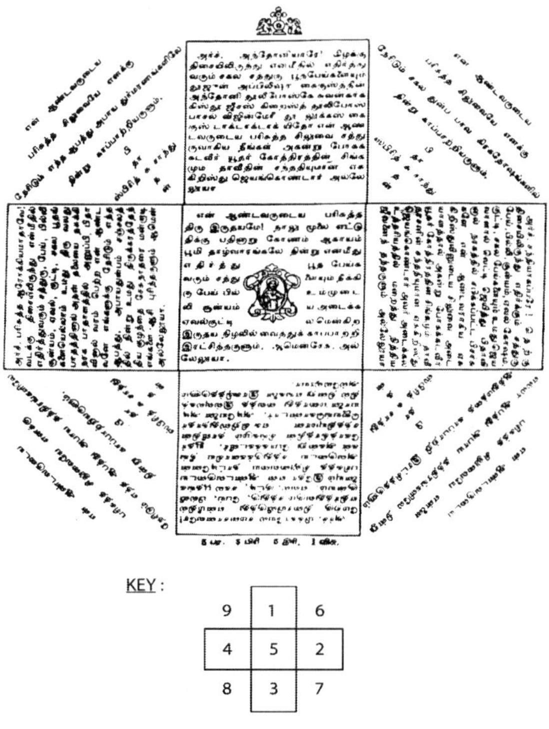

Fig. 7. *Satrusamharamala: The Mantras*

The other four directions in between are entrusted with the Holy Cross of the Lord. After "binding" the directions, each *mantra* ends by invoking the Holy Trinity.

The complexity of this *yantra* is noteworthy, and its full meaning is yet to be deciphered. Its origin also remains obscure. The close link of this *yantra* with the Indic tradition of regents of the eight cardinal points (*aṣṭadikpālakas*) is evident. Each cardinal point of the earth, according to traditional Indian geography, is being protected by a divine guardian or regent.[18] What is undeniable is that this yantra acts as a source of strength and protection to the Catholic fisherfolk in times of difficulty and danger.

Tantric Roots of *Mantravādam*

We may arrive at the following conclusions from the preceding discussion:

a. *Mantravādam* is practised widely by the fisherfolk parallel to their practice of the official Catholic sacraments and rituals.
b. Many Christian *mantras* are patterned after those existing in the Hindu tradition even when addressed to Christian saints using Christian symbols.
c. There are some Catholic practitioners of *mantravādam*, but many Catholics approach Hindu practitioners for their service.

A careful look into the cultural roots of the mantric tradition is necessary to understand the powerful hold the mantric rituals have on the fisherfolk. The mantric tradition in India goes back to ancient times. Vedic chanting was part of the *hōma* sacrifice for Vedic Indians. Here *mantra* referred to any passage in the Samhitā portions of the Vedas as distinct from the Brahmanas or the Upanisads. The term was also widely used in Buddhism as well as Jainism, and similar practices exist among the tribal groups in India.[19] Scholars today accept the existence of pre-vedic or even antivedic traditions of mantric formulas and rituals that were closely associated with Tantrism. As a result, Tantra itself is often called *mantra śāstra*.[20] The deep influence of the mantric tradition is

[18] Names of these regents vary. According to one list, Indra protects the east, Varuṇa the west, *Yama* the south, *Kuvera* the north, *Agni* the southeast, *Niruti* the southwest, *Vayu* the northwest, and *Isara* the northeast.

[19] See Agehananda Bharati, *The Tantric Tradition* (New York: Samuel Weiser Inc., 1975), chapter 5.

[20] See John Woodroffe, *Introduction to Tantra Sastra* (Madras: Ganesh and Co., n.d.), 84; N.N. Bhattacharyya, *History of the Tantric Religion* (New Delhi: Manohar, 1982), 16.

evident not only in the various Indic religious systems but also in the "imported" religions like Islam and Christianity.

A *mantra* is a sacred formula or a mystical verse addressed to a deity to acquire superhuman powers. It is composed of certain letters arranged in a definite sequence of sounds. The relationship between vowels and consonants (*vaṛṇa, nāda,* and *bindu)* in specified configurations manifest a particular deity. The central idea is that certain sounds uttered in the prescribed manner produce cosmic vibrations; this energy can be directed to serve specific purposes, good or evil, with the help of the deity. In a *mantra,* this cosmic energy is believed to reside in pure form. The efficacy of a *mantra* requires correct recitation according to proper rhythm (*svara*) and sound (*vaṛṇa*), and repetition of the same for a specified number of times and days. *Bīja* (seed) are short, unetymological vocables such as *Hṛīm, Kṛīm, Aim, Phaṭ,* etc., and each deity has her own *bīja* (eg. *Kṛīm*—for Kāḷī, *Hṛīm*—for Māya, *Rām*—for Agni, etc.)[21] The *bīja* is the microcosm of the deity and of the universe and is one of the seven limbs of a *mantra.*[22] A *mantra* is different from a prayer in the sense that in the latter, a worshipper chooses the words that would convey the real meaning of the prayer. But in a *mantra,* the real meaning is the *svarūpa* (own form) of the particular deity, which may not be explicit in the word.[23] In mantric rituals, the number 108 is particularly significant; if the incantations are to be fruitful, the string of beads *(māla)* used for the repetition of a *mantra* should have 108 beads.[24]

Keṭṭu or *digbandanam* (=binding of directions) is a central concept in mantric rituals; this refers to a preliminary meditation meant to ensure spiritual safety during the time of the ritual from supernatural beings believed to inhabit the four directions. *Mantras* are classified as masculine, feminine or neutral. Usually, masculine *mantras* end with *hum, phaṭ,* etc. Feminine *mantras* end with *tham* or *svaha.* Neutral ones end in *namah;* the neutral *mantras* are said to lack the force and vitality of the other two.[25] Based on the purpose for which it is applied, *mantras* are usual-

---

21 L.R. Chawdhri, *Secrets of Yantra, Mantra, and Tantra* (New Delhi: Sterling, 1992), 84.

22 The other limbs are Ṛṣi, Chandas, Kilāka, Śakti, Nyasa, and Dhyana. See Bhattacharyya, chapter 11.

23 Woodroffe, *Tantra Sastra,* 85.

24 Bharati, *Tantric Tradition,* 145.

25 See Sudhir Kakar, *Shamans, Mystics and Doctors: A Psychological Inquiry into India and its Healing Traditions* (New York: Alfred A. Knoff, 1982). Kakar tries to analyze the mantric syllables from a psychological point of view, and suggests that the *bījas* appear to be aural symbols for certain forms of inner psychological experience ranging from simple sensations to more complex emotional states.

ly discussed under six categories. These are: to heal (*śanti*), to seduce (*vaśyam*), to obstruct (*stambhanam*), to create enmity (*vidvesaṇam*), to expel (*ucāṭanam*), and to destroy (*māraṇam*). Each category has its own elaborate ritual system (*ācārās*), which are broadly considered under the "right" tradition (*dakṣiṇācāra)* and the "left" tradition (*vāmācāra).* Elaborate rituals consisting of the five M's (*pañcamakāra: madya*—wine, *māmsa*—flesh, *matsya*—fish, *mudra*—cereals, *maithuna*—sexual union) are specific to the left tradition.[26]

Mantras are closely related to and often used along with *yantras* (also called *chakras* or *mandala*). *Yantras* are of vital importance in the Tantric tradition, and consist of diagrams of geometrical patterns engraved or drawn on paper, metal, wood, stone, or any other substance, and used as religious/ mystical symbols. In worship, a *yantra* is an abstract image of the deity, and so acts as a substitute for the deity's iconographic image. The deity is invoked by reciting the specified *mantra* and following ritual prescriptions, and the *yantra* "becomes alive" with the living presence of the deity. Therefore, a *yantra* is a dynamic symbol, unlike other anthropomorphic images that remain static.[27] A *yantra*, being a linear configuration, has usually a center and several concentric primal shapes like triangles, squares, circles, hexagrams, or rings of lotus petals. There are specified positions and manner of writing the items in the *yantra.* Three methods usually followed are face to face (*sammukhom)*—for doing good, turned outward (*bahirmukhom)*—for causing harm, and written obliquely *(thiryangmukhom)*—for other purposes. Occult *yantras* serve as mystical diagrams of magical power and are used in healing sickness or as good-luck charms or for exorcisms or to prevent calamities. They are also worn as talisman or amulets for protection.

Both *mantra* and *yantra* are important ingredients of the Tantric tradition. What *mantra* does with sound, *yantra* does with visual images. The human body itself is a *yantra* in Tantrism; "that which is not in the body is not in the universe," is the Tantric maxim.[28] *Mantra* and *yantra*, to the Indian mind, form an effective bridge between the microcosm and the macrocosm, between individual consciousness and cosmic consciousness. The basic conceptual foundation of Tantrism lies in Śakti, i.e., the power of the Supreme Being conceived as the female principle through which the universe becomes manifest. Tantrism is often

---

26 For a detailed discussion on the *ācārās,* see Woodroffe, *Tantra Sastra*, 76–80. Bagcchi refers to three currents of Tantric tradition, *Dakṣiṇa, Vāma,* and *Mādhyama*; see Bagchi, "Evolution of the Tantras", in *Studies on the Tantras* (Calcutta: The Ramakrishna Institute of Culture, 1989), 14–18.

27 *Encyclopedia of Religion,* s.v. "Yantra".

28 Bhattacharyya, *Tantric Religion*, 7.

identified with Śaktism, a religious system that arose in the medieval period since Tantric ideas and practices were integral to it. In fact, Tantric ideas like the notion of Śakti as the Primordial Mother or Devī have penetrated all Indic religious streams including Buddhism. It is argued that Tantras have evolved as the philosophy/theology of the masses, especially the lower classes, stressing empirical knowledge and everyday concerns as against the "pure knowledge" of the Vedas.[29]

There existed a sect of Tantric followers known as *Kerala Sampradāya,* and it is proof enough of the geographical spread of Tantrism along the Kerala coast. Andre Padoux writes that Kerala was one of the main centers of Tantrism in ancient India.[30] The main facet of Śiva in Tantrism is that of Vairavan to whom the fisherfolk address many *mantras*.[31] The tradition of *Saptamatṛkal* (seven mother-goddesses) and that of Gaṇeśa (in whose name is the ruined temple at Gaṇapathi Vilākam) are believed to be closely related to Tantrism. These facts strongly support the view that Tantrism had played a vital role in shaping the neithal religiosity of the fisherfolk. This background is important in explaining the prevalence of mantric rituals and the centrality of Amma in the coastal religious tradition.

In conclusion, we may say that it is the complex world of rituals that sustain the everyday life of the fisherfolk, giving them strength and protection before all adversity. In the case of people on the margins, the cultural and religious horizons interpenetrate deeply in the realm of rituals. The ritual scenario illustrates that the fisherfolk retain many beliefs and practices of their pre-Christian past while affirming their Christian heritage earnestly. Many popular rituals have put on a clear Christian garb while staying deeply rooted in the neithal/Indic religious heritage. How do the Christian fisherfolk negotiate these diverse strands of religiosity in order to shape their sacred world? is a question for us to explore. The next chapter is an attempt in that direction.

29 Bhattacharyya, *Tantric Religion*, 3.

30 *Encyclopedia of Religion*, s.v. "Tantrism". See also Bharati, *Tantric Tradition*, 89.

31 Śiva in many of his epithets—*Rudra, Mahakala, Śambhu* etc.—is worshipped in the Buddhist Tantric and partly in Jaina Tantric traditions. See, Bharati, *Tantric Tradition*, 135.

CHAPTER 6

THE MOLDING OF NEITHAL CATHOLICISM

The Long Road of Assimilation and Synthesis

The Catholic history of the coastal people of south Kerala goes back to the period of their mass conversion to Catholicism through the work of the Portuguese missionaries who arrived four and a half centuries ago. Ever since, they have been scrupulously guarding their Catholic faith and identity. It is evident from their Christian names and cultic and devotional life centered primarily on the village church and the clergy. The skyline of most coastal villages of the district is marked by imposing churches and numerous shrines. But such external manifestations of a Christian identity may be deceptive as these may hide behind them features of their lived religion that are less visible. Our discussion in the earlier chapters indicates a superimposition of Iberian Catholicism over an Indic form of neithal religiosity that had nourished the community for ages. At a manifest level, the professed beliefs and sacramental practices seem to be no different from those of the universal Church. A more careful search is necessary to unravel the elements and dynamics of the lived religion of the fisherfolk.

The three basic components that go in to constitute a religion in the process of institutionalization are beliefs, rituals, and organization. Beliefs may be understood as patterns of ideas or conceptions that express the religious experience of the community either as systematic formulations in creeds or doctrines or through symbolic modes of expression. But for the fisherfolk, faith statements and creedal formulas are insignificant. Their true beliefs find expression in their numerous rituals of daily life and in the myths they live by. A proper grasp of such rituals and myths is inevitable to get a glimpse of their conception of matters divine. The term "divine" is preferred to God or the sacred. The term God, which originated as a common name, in course of history came to be associated with the Abrahamic "One, True God." With the spread of Christianity and

Islam, the term God acquired a hegemonic quality and began to reduce gods of other cultures to "mere gods." This background inhibits the use of the term "God" in an intercultural context.[1] The term "sacred" has been popular with the social scientists, and it is surely broader in conception. The difficulty becomes evident when it is used as the reverse of profane. Such a demarcation and apparent opposition between the sacred and the profane is quite alien to the cultural contexts of the fisherfolk. The term "divine" is preferred as its polysemic quality can do better justice to the conceptual world of neithal religiosity.

The essential background of the traditional neithal culture that characterizes the coastal people was discussed in chapter three. The discussion, based on the Sangam literature, highlighted how various Dravidian beliefs and divinities have contributed to the shaping of a culture and a religious universe that was specific to the coastal context. The predominance of the worship of the mother goddesses and of Murugan closely links the fisherfolk with the religion of ancient Tamiḷakam. Deities like Varuṇan, the god of the ocean, and Kumari, the virgin sea goddess, might have been added by a sea-centered people. Neithal religion tries to express the composite of this total phenomenon but within a broader frame of an Indic religious past.

There is sufficient evidence to believe that Buddhism was popular and widespread along the coastal region and among those outside the caste hierarchy. A major Buddhist center was Srimūlavāsam, now believed to be under the sea. Nagam Aiya, after discussing the remnants of Buddhism in Travancore, states that Brahmins have appropriated and adapted many Buddhist/Jain temples to convert them into Hindu temples; he gives examples of many such temples in south Kerala.[2] Buddhism began to decline by the eighth century and finally disappeared in the twelfth century under the aggressive onslaught of Vedic Hinduism, especially under Sankarācārya, the philosopher-sage. The Bhakti movement led by the Śaiva Nāyanmārs and Vaiṣṇavite Alwars also played its role in eliminating Buddhism and Jainism from among the masses. Many elements of these religions, at the same time, were appropriated by the popular form of Hinduism that flourished. The present-day expressions of worship, images, processions, and festivals in every religious tradition including the Christian are proof of this legacy. Dravidian deities were transformed and got assimilated into the Hindu pantheon under the influence of Vedic Hinduism. Vaidyanathan writes: "Thus the Dravidian goddess Koṛṛavai became Durgā, Kālī, and Bhagavathy. The popular

1 See *Encyclopedia of Religion*, s.v. "Deity".

2 Nagam Aiya, *The Travancore State Manual*, Vol.2, 244.

non-Aryan deity Śastha came to be looked upon as Hariharaputra, the son of Vishnu (Hari) and Śiva (Hara), in order to make him acceptable not only to the Śaivite and Vaiṣṇavite sections of the Hindu population but also to the converts to Hinduism from within the Buddhist fold."[3]

We may rightly conclude from the above discussion that the religious face of the coastal people at the time of their conversion to Christianity was that of popular Hinduism, but one that resulted from a complex process of assimilation and synthesis between various strands of Indic religiosity. It was into this mold of Indic religiosity that the Iberian Catholicism got implanted in the sixteenth century, thus initiating a new process of assimilation and synthesis. We shall discuss below some significant themes that have been central to this process.

Cintāthira Māta and the Mother Goddess Tradition

Coastal people use terms like Amma and Māta frequently in their daily life, whether the context is religious or not. The word Amma in the Tamil lexicon is rich and evocative and may be translated as mother, mistress, or lady. This is a term of respect that can be used for any person of the female gender, whether human or divine. The traditional mother goddess figures are called Amma or Amman. The sea also is reverentially called Amma or Kadalamma (Mother Ocean). But for the fisherfolk of Vizhinjam and its neighborhood, Amma is primarily Cintāthira Māta, the patron of the community and of the church. She is accorded a prime position in the social and cultic life of the people.

Cintāthira Māta is the colloquial form of *Sindhu-yātra-māta*. The term *Sindhu* has the following meanings in order of priority: (1) the ocean, (2) water (3) river in general, and (4) the river Indus.[4] In the present context, *Sindhu* means the ocean and *yātra* means journey. Thus, she is considered the Lady of Good Voyage, the protector of seafarers. Her statue adorns the main altar in both the churches and atop the imposing bell tower of the new church at Vizhinjam.[5] The present statue looks very similar to that of Our Lady of Velankanni, except that she holds a ship in her right hand in place of a sceptre. She is described as the heavenly mother, "wearing the sun as her apparel, the moon as her footstool, and

3 K.R. Vaidyanathan, *Temples and Legends of Kerala* (Bombay: Bharatiya Vidya Bhavan, 1988), 6.

4 *Tamil Lexicon*, Vol. III, University of Madras, 1982.

5 See R. Nagaswami, *Tantric Cult of South India* (Delhi: Agam Kala Prakashan, 1982), 19. He refers to *Cintādevī* in *Maṇimekalai*, the well-known Tamil epic of Buddhist origin. *Cintādevī* is one of the four forms of the goddess, according to the Śakti cult prevalent in Tamiḻakam. But this is not proof enough to show any vital link between *Cintādevī* and Cintāthira Māta.

twelve stars as her crown, and holding the child in her left arm and a ship in her right arm, looking up with her right eye towards heaven and focussing her left eye down on the earth."[6] The original statue, made of wood, depicting her as seated in a chair, was smaller in size and is now lost. The legend has it that the statue landed at Vizhinjam in a wooden box; it had a note in it asking the finder to contact the sender for putting up a church at the landing spot. Some say that other villages had tried in vain to catch hold of the statue since it turned its back as they approached it. It was only at Vizhinjam that the statue stayed in a proper position. A new church was built as instructed in the place of the old thatched shed and the statue was installed in it.[7]

Image 4. The image of Cintāthira Māta

The cult of Cintāthira Māta reaches its climax on the day of the parish feast that falls in the month of January. It is a ten-day-long village festival that is eagerly awaited by all. The statue of the Māta, richly decorated and illuminated, is taken around the village in procession on the ninth day.

6 A.I. Antony, "Vizhinjam Parish through History," an unpublished write-up in Malayalam.

7 The cult of Cintāthira Māta does not seem to be confined to Vizhinjam. In Sri Lanka, there is a 120-years-old church dedicated to Cintāthira Māta situated at Amrutakazhy in Batticaloa on the East coast; it is a sub-station of Puliadikuda parish. The statue of the Madonna is placed in a ship and is locally called *Kappalēri Māta* (the Māta who has boarded the ship). The legend has it that a Portuguese ship, caught in a storm in the sea, was brought ashore by Virgin Mary miraculously.

The procession starts from the old church on the seashore and ends in the new church. Recently, the route has been shortened due to the existing situation of intercommunal tension. While joining in the procession, many people prefer to dress up in the traditional costume of the pious association *(sabha)* they belong to as an act of devotion. The order and position of each group in the procession is determined by certain conventions. The ancient confraternities *(compiri sabhakal)* always get precedence over the others and stay close to the statue. All along the route, devotees throng to make their offerings and to receive the divine blessing *(prasādam)* in the form of pepper and salt.[8] The procession is followed by elaborate cultural program like drama or musical concert held on an open air stage on the beach. The statue of the Māta is placed in the new church for a few days for veneration before it is taken back to the old church.

Cintāthira Māta has come to be the protector and refuge of the fisherfolk of the village. "It is her power that sustains us," says the catechist *(kaṇakkapiḷḷa)* of the church. It is she who protects them in calamity—whether in the sea or in the marketplace or in the event of a communal clash. The communal conflict of 1995 that claimed the lives of four Catholic fishermen and two Muslim fishermen is still fresh in the people's mind. Tharsilamma, an elderly fisherwoman, is convinced that it was Amma who was fighting for them, standing between the two warring groups and blocking the bombs thrown at them. "Our Amma fights for us," is a refrain often heard. When a fisherman is reported missing in the sea, or when a calamity occurs in the village, women and children rush to the statue in the old church and cry aloud in frenzy, "You come down, Amma! You come with your son and save us!" Fear does not prevent the menfolk from venturing into the roaring waves and into the deep sea, for they know Amma would protect them. Women call upon Amma and pray for the protection of their husbands and sons; "Amma, *kappalkkāri* (ship-bearer), who lifted up the sinking ship, please protect our children!"

Thurston writes that the caste deity of the fisherfolk was goddess Bhadrakāḷī, represented by a log of wood placed in a hut called the temple, and to whom cock-sacrifice and fruit offerings were made four times a year.[9] Studies show that the cult of Devī was an ancient Dravidian practice prevalent along the Kerala coast. According to *Keraḷolpathi*, Paraśurāma reclaimed the land of Kerala from the sea and established

---

8 Both pepper and salt are counted among the "eight poisons" (*aṣtaviṣam*), according to Hindu mantric texts. The offering of these substances in *mantravādam* is closely associated with rituals aimed at the destruction of one's enemy (*śātrusamhāra*).

9 Thurston and Rangachari, *Castes and Tribes*, 106.

sixty four settlements (*gramams*), the goddess Durgā was set to guard the seashore on the west, and the god Śastha the ghats on the east.[10] Paraśurāma also established 108 images of the Durgā Devī on the seashore, and also erected shrines for serpents and petty gods.[11] This is further corroborated by the existence of numerous ancient Bhagavathy (popular name for Bhadrakāḷī or Kāḷī) temples all along the Kerala coast. Tradition has it that there existed seven temples in the tiny village of Vizhinjam from ancient times, although many of these are nonexistent today. The Bhagavathy Amman temple at Pulloorkonam is a protected monument under the Archaeology Department. It is built fully of granite slabs in the Chola style, and dated back to ninth century CE, and is dedicated to Saptamātṛkaḷ (seven mother goddesses), linking the coast strongly with the mother goddess tradition of ancient Tamiḷakam. The dilapidated temple at Gaṇapathy Vilakam too was a Devī temple.

Bhadrakāḷī or Kāḷī is usually depicted as a grotesque female figure with terrifying features. David Kinsley describes her features thus:

> She is always black or dark, is usually naked, and has long dishevelled hair. She is adorned with severed arms as a girdle, freshly cut heads as a necklace, children's corpses as earrings, and serpents as bracelets. She has long, sharp fangs, is often depicted as having clawlike hands with long nails and is often said to have blood smeared on her lips... . She is usually shown on the battlefield, where she is a furious combatant who gets drunk on the hot blood of her victims, or in a cremation ground, where she sits on a corpse surrounded by jackals and goblins.[12]

Scholars say Kāḷī is associated with the periphery of the Hindu society as she is worshipped by tribal or low caste people. "Indian mother-goddess temples are a direct growth from primitive tribal cults, each of local origin, later brahminized", states Kosambi.[13] Thus she was made one of the manifestations of the Mahādevī, the Great Goddess in the Indian tradition. The widespread cult of mother goddesses has been subjected to many studies. Svelebil reconstructs the archetype of these as Ūramma, the protective mother goddess of the *Ūr*, the village settlement.

---

10 *Keraḷolpathi* (Origin of Kerala) is an ancient collection of stories and legends about the origin of the land of Kerala written in Malayalam. Its author is believed to be Thunchathu Ezhuthachan, a seventeenth century scholar, but the manuscript was later discovered by Herman Gundert and published in 1868 from Mangalore.

11 Nagam Aiya, *State Manual*, Vol.1, 217.

12 David Kinsley, *Hindu Goddesses: Visions of the Divine Feminine in the Hindu Religious Tradition* (Delhi: Motilal Banarsidass, 1987), 116.

13 D.D. Kosambi, *Myth and Reality: Studies in the Formation of Indian Culture* (Bombay: Popular Prakashan, 1962). 68. See also Kinsley, *Hindu Goddesses*, 117.

> Ūramma was the archetype of all the *ammas* in the Dravidian speaking south, connected with agricultural operations, rain and fertility rites on the one hand, and with epidemic diseases on the other hand... . Ūramma was most probably the Neolithic prototype of the ubiquitous Goddess or Devī in her many various multiforms... .[14]

It is probable that the group of goddesses known as Saptamātṛkaḷ (seven mothers) represent the many village goddesses worshipped by the common people as their protective deity.[15] In the Tamil tradition, they are not really mothers but virgins or maidens (seven *kannimar*). The dualistic opposition between motherhood and virginity does not seem to be a major concern to the Dravidian mind. The link between crossroads and mother goddesses is well documented. "The crossways are logically the original sites for the mother goddess cults," argues Kosambi with the help of field data.[16] The significance of crossroads in *kaṇṇupedal* rituals supports the deep influence of the mother goddess tradition on the belief system of the fisherfolk.

Belief in a mother goddess appears to be a universal phenomenon seen among all peoples irrespective of color or continent. However, studies show that "no civilization in the world developed goddess worship so elaborately as did India."[17] What would distinguish the goddess or Devī image in India is the aspect of ambivalence; both benign and fearsome features (*śāntabhāva* and *rudrabhāva*) coexist within the very same image. In her benign form, Devī "represents the world as unceasingly fruitful, full of awesome energy that pervades and nourishes all creatures."[18] As a mother, she showers on her children all the bounties at her disposal. Fertility, material abundance, food and nourishment, the embodiment of female beauty and wisdom are associated with the benign form. Divine figures like Lakṣmī, Pārvatī, and Pṛthvī are manifestations of this form. Her fierce or fearsome forms are associated with war, blood, destruction, and death. In her protective role, she often assumes the form of a fierce warrior. Her weapons are mainly the protruding teeth and sharp nails. Her taste for blood is the reason why she is often worshipped with blood

---

14 Zvelebil, *Tiru Murugan*, 6.

15 The number and names of the *mātṛkaḷ* vary according to each tradition. *Devī Māhātmyam*, a work of the medieval period refers to seven Śaktis, closely resembling the male gods. But this depiction is arbitrary and betrays a pure Brahmanical portrayal. See Kinsley, *Hindu Goddesses*, 156,

16 Kosambi, *Myth and Reality*, 95–96; See also Kinsley, *Hindu Goddesses*, 155, and James Freeman, *Untouchables: An Indian Life History* (New Delhi: Indus, 1993), 403.

17 *Encyclopedia of Religion*, "Goddess Worship—An Overview".

18 Kinsley, *Hindu Goddesses*, 149.

sacrifices. Divine figures like Kaḷi, Camunda, and Mariamman are usually the fearsome forms of the goddess.[19]

Kāḷī is the embodiment of her fierce nature (*rudrabhāva)* but not apart from her benign nature *(śāntabhāva)*. The two are closely interrelated, and each nature requires the existence of the other. In some texts like *Karpuradistotra*, Kāḷī is portrayed as young and beautiful, having a smiling face, and making gestures that dispel fear and offer boons.[20] The benign dimension is emphasized by the devotional tradition while the fearsome nature is the focus in the Tantric tradition. In the evolution of the Devī cult, the two paradoxical features went through noteworthy changes. The long interaction between the Buddhist and Hindu Tantric schools led to its culmination in the concept of Devī in ten different forms or aspects, including the benign and the fearsome. But devotees never experience the distinction as a paradox.[21]

Image 5. Cintāthira Māta atop the church

19 See V. Subramaniam, "Stages and Paradoxes in the Evolution of the Supreme Hindu Mother Goddess", *Mother Goddess and Other Goddesses*, ed. V. Subramaniam (Delhi: Ajanta Publications, 1993), 5–16. It gives a historical sketch on the evolution of the benign and the fierce features of the Devī in the Indian traditions. He refers to the Purana, *Devī Māhātmyam*, which depicts Pārvatī splitting herself into the fierce Kāḷī and the bewitching Ambikā, in order to kill the demon Mahiṣasura who threatened the Devas. She is fierce to the demons but benign to the Devas. A comparable tradition is present in southern India in the story of Kovalan and Kaṇṇaki as narrated in the Tamil epic, *Śilapathikāram*.

20 Kinsley, *Hindu Goddesses*, 125.

21 See V. Subramaniam, "Stages and Paradoxes," 11.

We can easily infer from the above analysis that the notion of the divine feminine, the Devī, is central to the conception of the divine in coastal worldview. Other studies too support this inference.[22] The long tradition of the worship of a mother goddess, a feminine Śakti, has left very strong marks on the religious consciousness of the fisherfolk that couldn't be erased even with the best missionary efforts. This not only survives but also continues to influence the everyday life of the coastal Catholics although some elements have been either deliberately destroyed or conveniently forgotten. The Nuḷayas (Hindu fisherfolk) residing in some other villages like Panathura and Kovalam confirm that their ancestors had run away from Vizhinjam to escape the mass conversion drive in the sixteenth century. Missionary documents testify that systematic attempts had been made to demolish the "pagan" temples and images during this period. It was common practice in missionary history everywhere to force the neophytes to make a total break with their "pagan" past and to submit totally to the new faith, although attempts were not always successful.

Studies on conversion show that change of faith is not at all a simple affair. It may be easy to replace temples and idols with churches and statues, but replacing the meaning structure of a culture and the mental images of a people are far more difficult. These persist even after conversion in diverse forms, often outside the precincts of the official religion. In the case of the Catholic fisherfolk, Amma remained ever alive in the collective religious psyche of the community. *Kenkādeviamma* would never fully vanish from their life, nor would Bhadrakāḷī. Our Lady of the Assumption (Paralōka Māta) was superimposed on the neithal religion. But four and a half centuries of religiocultural osmosis have given birth to Cintāthira Māta as the new symbol of their religious consciousness and identity. She appears to synthesize harmoniously the diverse forms and features of the image of Amma, in a manner quite acceptable within the parameters of the Catholic religious tradition.

Warrior Divinities and Christian Warrior Saints

"Our life is a life-long wrestle with the ocean," these words of Markos sum up the coastal people's view of life and its mysteries. Wrestling, fighting, and struggling are words that recur in their dialogue. There seems to be an irresistible urge for religious symbols and expressions that are attuned to a militant mindset, and it is vividly expressed in their preference for warrior divinities. The religious world they construct cannot

22 See Kalpana Ram, *Mukkuvar Women*, chapter 3 and 4.

stay unrelated to the neithal culture in general and the hunting ethos of the fisherfolk in particular.

The cult of warrior gods has deep roots in the culture of ancient Tamiḷakam that included the coastal regions of Thiruvananthapuram. One cannot miss the influence of Murugan or of Vairavan (Bhairavan), two ancient Dravidian warrior gods, on the coastal community. The ritual of *eilamidal* would reveal the residual religious beliefs related to Vēlmurugan.In the folk song they sing while pulling the shoreseine net, the chorus has the word *Ovalē* repeated often. "*Ovalē* is a corrupt form of '*O Vēla*!' which is an invocation of Vēlan that is Vēlmurugan," the wise elders explain. Data gathered from the nearby Hindu fishing villages show that in their version of *eilamidal*, the song is addressed explicitly to Vēlmurugan mentioning his features like "the one holding a *vēl* (spear) in his palm." The Catholic fishermen invoke Vēlan in their occupational songs, and this indicates that remnants of Murugan worship, part of their Nuḷaya heritage, are still retained.

Murugan (Vēlan) is a Dravidian god, popular all over Tamil Nadu, southern Kerala, and Sri Lanka. The Sangam literature depicts him as the protective god of the mountainous region; references to him and being possessed by him are frequent in the *akam* poems. He is referred to as "the Radiant/Red one" or Cēyon or Cevvel (red *Vēl*). In all probability, he might have been a tribal-hunter god, who later evolved into a warrior god in the later Sangam works. Sanskritic influence led to his metamorphosis into the complex deity of Skanda-Subramanya-Murugan as he is known today.[23] The original Dravidian Murugan was the son of Koṟṟavai, the mother goddess of war and victory, though, in the later myths, he came to be the son of Śiva. He is the symbol of heroism, and so is depicted as a blazing warrior adorned with a shiny spear, and riding a peacock.

In the Catholic tradition, it is easy to see many of these warrior qualities transferred to many Christian saints. In spite of her ambivalent nature, the Māta is predominantly a benevolent mother; this is partly because saints like St. Michael, St. James, and St. Antony take upon themselves much of the warrior qualities of Indic divinities, thus freeing the Māta to be more a mother. St. Michael in the Christian tradition is the first of the seven archangels, leader of the Hosts of Heaven who defeated Lucifer when he revolted against God. He is depicted as a winged warrior because he fights against the powers of darkness. In Daniel's vision, he appears as the great prince who defended the people of Israel (Dan 10:13, 21; 12:1 NRSV). In the Book of Revelation (Rev 12:7–9), St. Michael is

23 Zvelebil V. Kamil, *Tiru Murugan*, 7–8; Srinivas Iyengar, *History of the Tamils*, 556 & 612.

the leader of the Angelic Host that defeated the dragon who is shown at his feet. The sword of St. Michael has clear reference to Genesis that says, "At the east of the Garden of Eden he placed the Cherubim and a flaming sword which turned every way, to guard the way to the tree of life" (Gen 3:2 NRSV). An allusion to this sword is made in the ritual of the Wheel-Seat of St. Michael.

Santiyagappar (Santiago or St. James) is an equally popular warrior saint. He is in fact St. James the Great, the son of Zebedee, a fisherman, and a disciple of Jesus (Matt 4:21–22); but the Gospel account shows little trace of a warrior figure, except the vague reference to "sons of thunder" (the name Jesus gave to James and his brother John, as recorded in Matt. 3:17), and of having "fiery disposition" (Luke 9:52–55). But in Tamil, he is called *Pataimiraṭṭi Santiyagappar* (St. James who repelled the army). This may have reference to a Spanish legend (inherited by the fisherfolk through Iberian Catholicism), according to which the tomb of St. James was located in a place called Compostella in Spain in the ninth century. As a result, St. James was declared patron of Spain in the fight to reconquer the land from the Moors. In support of the Christians, the saint is believed to have appeared with a sword in hand on a white horse, on the eve of the battle of Clavijo in 930 in which the Christians won; hence the name, Santiago Matamoro (killer of Moors).

St. Sebastian, another popular warrior saint, was a soldier-martyr of the Roman Empire under Diocletian. His devotion was carried by the Portuguese military for motivating their soldiers to die for the empire, convinced that they were dying for faith as well.[24] St. Antony of Padua has no warrior background, being a Franciscan friar who became a renowned preacher. He is usually invoked for the restoration of lost property. But for the fisherfolk, St. Antony exhibits many features of a fiery deity who is invoked in *mantras* to dispel the enemies or "to cut them to pieces." He is referred to as *muni* (sage) in some *mantras*. The predominant place accorded to these warrior figures in the rituals and practices of the community indicates their attempt to discover meaning in the rich Christian tradition itself without losing their grounding in the Indic worldview.

The Coexistence of the Catholic and the Indic

The analysis of Cintāthira Māta and the warrior saints throws light on some key elements in the coastal conception of the divine. This conception portrays features of the divine world that belong to two distinct religious traditions—that of Iberian Catholicism and that of Indic religiosity.The two traditions seem to coexist under what is seen as "Neithal

[24] Teotonio de Souza, *Goa to Me* (New Delhi: Concept Publishing Co., 1994), 96.

Catholicism." Many beliefs that underlie rituals of their Catholic life often show a close resemblance to rituals in the Indic tradition. Official prohibition or sanction does not deter them since the Indic rituals had become part of their everyday existence over generations. Their practice of Catholicism as taught by its creeds and precepts goes on simultaneously. The coexistence of both the Indic and the Catholic in the midst of an apparent conflict between the strong emotive appeal on the one hand and the official disapproval by the church on the other is a fact to reckon with.

How does the community deal with this problem of incongruity between two divergent religious traditions? Does it create inner conflict and a feeling of guilt in the fisherfolk? Indications are that the two streams blend well in some situations, although at other times these run on parallel lines or lead to conflict. The aggressive attempt of the early missionaries to eliminate whatever belonged to the Indic traditions forced the fisherfolk to retain the same at a latent level of religious practice. More than retaining, they were trying to reconcile the two outside the purview of church controls. That the community has survived four and a half centuries of existence in a predominantly non-Christian milieu with a strong Catholic identity is proof enough of the community's resilience and relative success in reconciling the two. It is also evident that the community is not unduly perturbed even if some elements are not reconciled; ambivalence is part of their cultural ethos, and also of their lived religion.

Kalpana Ram has analyzed Mukkuva Catholicism on the basis of the Manichean principle of the opposition of good and evil supernatural beings.[25] Mary, the Māta, is posited on the side of good assisted by the saints, while on the other side, the side of evil is the Hindu goddess Issakai and her demon companions like Mādan, and various ghosts (*āvi*) and spirits (*peiy*). Since the Hindu goddess is viewed as "evil incarnate," as the personification of *aṇanku*, the "destructive and capricious power in the world," the battle between good and evil is conceptualized as a battle between a Christian God and a Hindu goddess, leading to a basic antagonism between the two communities. This reading of Ram is not in consonance with the data available from the Thiruvananthapuram coast. To depict the relationship between the Hindu divinities and the Christian Māta and saints as one of antagonism between good and evil is too simplistic. The beliefs and attitudes of coastal Catholics, as evident from their rituals and practices, reveal a different framework.

The Catholic fisherfolk go to the Hindu temple at Thaṭṭatheru (goldsmith's street) and make offerings in order to get back lost articles. They believe that the Hindu goddess is more prompt in such cases. "Their god

[25] Kalpana Ram, *Mukkuvar Women*, 63–67.

pays the wages on the spot; our god moves slowly," is a popular saying. "Our God is a god who asks us to forgive seventy times. Their gods will respond to requests more swiftly. In a sense they are more powerful; for, are they not angels cursed to be devils, and thus possessing the powers of both angels and devils in them?"—these words of a fishvendor woman seem to portray a nuanced understanding of the divine world where both the good and the bad have a place. The same logic explains their unreserved use of mantric rituals that invoke Bhairavan, a Hindu god, in order to ensure a good catch. Practical concerns of daily bread and fish seem to take precedence over abstract thoughts about good and evil. What brings goodness and fortune is perceived as good.

A community's notion of the divine is closely associated with its understanding of good and evil. For, understanding evil in everyday life and dealing with it or warding it off is an existential concern in the risk-ridden life context of the fisherfolk. A study of coastal rituals indicates that a full picture of their conception of the divine will be available only after a study of their demonology as well. But demonology is a field beyond the scope of the present book. All through this process of interaction and even conflict, a silent process of osmosis and synthesis has been taking place. The invention of the Plate of St. Antony and the Wheel-seat of St. Michael are good illustrations of a process in which the religious trappings of the neithal past are given a specific Christian form and flavor. This has helped the community for centuries to withstand the threat of subtle appropriation by the dominant Hindu culture, and to prevent a total marginalization by mainline Catholicism, and thus to affirm their undeniable coastal/neithal Catholic identity.

CHAPTER 7

NEITHAL CATHOLICISM AS LIVED RELIGION

The earlier chapters discussed the phenomenon of coastal Catholicism with the aim of articulating the major contours of its sacred world. The coastal settlements and the seascape, the mysteries of the sea and the sky, the never-failing bounty of Mother Ocean, and the risk of an uncertain future, all play a part in this sacred world. The rich array of coastal rituals and the conception of a divine provider that is both benign and terrible become intelligible only from within this world. At the same time, the conversion experience initiated a new process of encounter between two divergent religious traditions giving birth to what may be called Neithal Catholicism as it is vibrant today along the Kerala coast. This chapter tries to explore some basic questions that emerge from this phenomenon and to draw its implications for a Christian theology that is truly rooted in the Indian soil.

Lived Religion—Coastal Catholics Become Real Subjects

The latent mode of coastal religion may appear an admixture of elements from various traditions with no inherent logic or method. But a careful reading would reveal craftsmanship at work, weaving together these disparate elements into a unified whole. Samuel Rayan writes: "All significant human activity implies faith (or unfaith), and therefore a theology (or a mammanology, a satanology, a demonology). And all theology has a method; it has its own tools and procedures, and its proper 'manner of grasping reality'. A method is to theology what grammar is to language."[1] Therefore, every religion at the latent level implies a latent theology and a proper method, and coastal Catholicism is no exception.

Human experience is the basis of all religion and theology, and every experience is context-bound. The primary experience of the fisherfolk, for generations, has been intertwined with their life and work in close

1 Samuel Rayan, "The How of Third World Theologies", ed. J. Russell Chandran, *Third World Theologies in Dialogue* (EATWOT, 1991), 45.

association with the ocean and the seashore. A unique neithal culture, with its specific worldview and religious conceptions, can be seen as the outcome of this long association. This neithal religiosity has been the strength and protection of the community in the face of the risk-ridden occupation they were engaged in. Other Indic religious traditions such as Buddhist, Tantric and popular Hindu have contributed to the molding of neithal religion at various stages of its development. But the experience of mass conversion to Catholicism by the mid-sixteenth century was a turning point in the life of the fisherfolk. They found themselves confronted by an alien religious system and worldview. The newfound religion was welcomed as a source of a new identity, but they had also to deal with an inner process of apparent incongruity. It was an encounter between two divergent religious traditions, viz. that of Iberian Catholicism and the pre-existent neithal religion. The attempt to shape what may be called coastal Catholicism has to be seen against this interaction. Their attempt was naturally confined to the latent plane, and the outcome was Neithal Catholicism, the lived religion of the fisherfolk.

To be on the seashore, geographically and socially, is to be on the margins of the mainstream society. This marginality extends to the realm of religion too. The latent mode of Neithal Catholicism has always remained on the margins of mainline church life, and this inherent marginality has contributed to the unique features that give the religion a flavor of its own. It is on the seashore that the fisherfolk become their true self and masters of their religious practice. Their concerns and questions change as the viewpoint shifts from the center to the margins. The fascinating yet frightening world of the ocean with all its treasures, the longing for protection from evil forces, and the urge to celebrate life raise ever new questions. The official mode of liturgical practices and a cerebral form of theology hardly offer space to deal with these questions. The view from the margins would reveal a story that is very different from what has been officially presented. The obvious example is the observation of the missionaries that the fisherfolk are "troublesome," "hard to deal with," "not very obedient to the church," and "resisting the authority of the priests."[2] What made the fisherfolk to disobey and to rebel was never understood nor looked into by the missionaries. The same view from the center continued during the period of Goan hegemony and continues to color pastoral approaches even today. As a result, official Catholicism reduced the fisherfolk to the level of mere consumers during the past four and a half centuries of their history—as masses to be converted, as neophytes to be instructed, and as Christians to be managed.

2 Joseph Thekkedath, *History of Christianity in India*, Vol. II (Catholic History Association, India, 1982), 186.

However, a process of silent transformation has been going on simultaneously at the level of lived Catholicism. This reshaping of their religion enabled them to effectively deal with the incongruities that emerged from the fusion of two religious traditions with divergent worldviews. All through its Christian history, the fishing community made efforts to assert as real subjects in the practice of religion both as of its creators and as its interpreters. First, they dared to reread and reformulate the Christian Scriptures, practices, and doctrines to suit their needs and their context. Second, they brought along elements of their neithal religiosity into their practice of Catholicism. In the former, they were interpreters of a tradition, while in the latter they became original creators. If the term theology can be freed from the hegemony of professionals, then the fisherfolk could be counted as theologians at the primary level.[3] This reshaping was at the same time a subtle form of protest and rebellion against the official form of Catholicism that failed to recognize their neithal worldview. Religion always involves power. Official religion tried to keep power away from the margins or to deny it altogether. But their Catholicism as lived made them subjects of a religion that empowers and liberates them. The radical contribution of lived religion to a people on the margins becomes evident here.[4]

As subjects of their own religion, the fisherfolk seem to be evolving an integral approach to religious history and to Scripture. Religion does not occupy a corner of their daily life or a certain period of the community's history; it permeates every space of their life and every period of their existence. In their practice of Neithal Catholicism, the fisherfolk give due recognition not only to their Christian history but also to their pre-Christian past. Some of the ritual practices are indicative of the rich storehouse of faith heritage lying dormant in the collective memory of the community. This collective memory is their history, and it is through this history that their theology has evolved over generations. This gives a new meaning to their understanding of Scripture and tradition. Their experience of encountering the divine is not confined to biblical books or to a period labelled as "history of the Church." As a result, their pre-Christian past becomes their "Old Testament" that is inevitable in illumining the New Testament. This integral approach to religious history and to Scripture has the inherent power to free people from the burden

---

3 Cf. Felix Wilfred, "Towards a Subaltern Hermeneutics", *Jeevadhara* 26, no. 151 (January, 1996), 56–59. In a subaltern hermeneutical perspective, the question "Who interprets?" becomes pivotal.

4 This does not deny the remarkable capacity Catholicism has displayed in the past to absorb popular religious practices, particularly in the veneration of saints. See P. Brown, *The Cult of the Saints* (London: S.C.M. Press, 1981).

of a conflict of religious traditions. Obviously, it may raise basic questions related to salvation history and the meaning of Christian tradition with far-reaching implications.

Ensuring Structural Identity and Meaning Frame

For people to perceive themselves as subjects of religion and theology is to recognize the specificity of their culture in the sacred realm. Two processes of interaction are noteworthy here: one, between the community and their neithal life context, and the other, between the two divergent religious traditions. It is the latter that is the main concern of this section.

Every religious tradition has evolved, and continues to function, within a specific and unique worldview. The missionaries brought with them not only the Catholic beliefs and practices but also a worldview that is specific to the Iberian Peninsula. The resultant encounter between the two divergent religious traditions implied more than a mere acceptance or rejection of certain articles of faith in isolation. Every article of faith is interwoven into the worldview of the people in question. Written documents mostly reflect only the missionary perspective that is basically Iberian. The inner dynamics as seen from the viewpoint of the fisherfolk, though scanty in written form, becomes vital to this analysis. For them, what happened was an encounter and dialogue between two worldviews.

Robert Schreiter, while analyzing the encounter between two religious traditions, utilizes the concepts of syncretism and dual religious system. The category of dual religious system is found helpful in unfolding the dynamics of the process in the present case, though it is used with reservation.[5] "Syncretism," he writes, "has to do with the mixing of elements of two religious systems to the point where at least one, if not both, of the systems loses basic structure and identity." The result is the birth of a new system out of this process.[6] A dual religious system is slightly different. According to Schreiter, "In dual systems, a people follow the religious practices of two distinct systems. The two systems are kept discrete; they can operate side by side. Sometimes one system is followed more faithfully than the other... in other instances the two systems may be followed

[5] For a detailed discussion of these concepts, see Robert J. Schreiter, *Constructing Local Theologies* (New York: Orbis, 1985), chapter 6. My reservation concerns the term "system"; it is too preposterous a term to make sense for the fisherfolk. System building has been alien to their practice of religion that is better understood as a "way of life"; it can hardly be fitted into any system.

[6] Cf. Louis J. Luzbetak, *The Church and Cultures* (New York: Orbis, 1988), 360. Luzbetak makes a distinction between syncretism as used in anthropology and in missiology. In the latter, it refers to "a combination of beliefs and practices that are theologically untenable."

almost equally."[7] In the history of Christian conversion, the emergence of dual religious systems has been one of the consequences whenever a rigid line was taken on the question of any cultural accommodation. The ruthless eradication of "pagan" images and practices and symbols of the fisherfolk by Francis Xavier and his associates is a case in point. The parallel existence of the Christian and the neithal religious traditions among coastal Catholics today is illustrative of a "dual religious system."

Schreiter discusses three kinds of dual religious system. In the first kind, Christianity and another tradition operate side by side. People follow both sets of rituals without seeing any contradiction. In the second kind, Christianity is practiced in its integrity, and selected elements from another system are also practiced. This dual practice is more evident in times of distress. However, the two are perceived incompatible or contradictory by the official church leadership, but not necessarily by the members. The third kind is called "double belonging" and specifically refers to situations where belonging to a religious tradition is closely linked up with a national culture, as in Myanmar or Thailand.[8] The coastal practice of religion seems closer to the second kind described above, since practice of Catholicism remains primary while useful practices from the pre-existing neithal traditions go on unhindered.

The notion of dual religious system, though helpful, does not fully explain the coastal experience. The two religious streams, at the latent level, need not always go parallel; there are also points of close interaction often leading to assimilation or integration. It is here that the creative religious mind of the fisherfolk operates as it refuses total appropriation by the dominant stream and molds a new identity in the process. In this creative synthesis, the fisherfolk make sure that its basic needs are met by these two traditions. They value their Catholic identity, since historically it has given them social status in an otherwise marginal context. They had only to gain socially by belonging to the Catholic fold. Hence, they carefully maintain the rituals and administrative structures of the church. In sociological terms, the Catholic tradition has contributed to the "structural identity" of the community. At the same time, they carefully preserve their Indic/neithal heritage to uphold a "meaning frame" as their reliance on pre-Christian rituals testifies. In their search for meaning, the fisherfolk still fall back on the Indic/neithal religious heritage and worldview. The Catholic tradition does not seem to have contributed significantly to alter that.

---

7 Schreiter, *Local Theologies*, 145.

8 Schreiter, *Local Theologies*, 148–149.

Is there a contradiction between the identity structures and the meaning frame in their everyday life, especially when the official church is keen to suppress the neithal heritage? Field inquiry has revealed that the fisherfolk, in general, do not view the two as contradictory. A different religious logic is at work here: the religious boundary of the fisherfolk, rooted as it is in a neithal worldview, is not a closed one since religion is not experienced in exclusivist terms. The boundaries remain open as built on a conjunctive logic, i.e., *both* this *and* that. In contrast, the inherent logic of the Christian tradition has been a disjunctive one, i.e., *either* this *or* that, but never both.[9] In this exclusivist thought pattern, if you are a Christian, you cannot be any other. That has been the ruling logic of the Western church, finding its culmination in the maxim, "no salvation outside the church." This intolerance justified every attempt to eliminate all pre-Christian religiosity and prevented all attempts at a creative synthesis. Creative encounter and dialogue existed only within their lived Catholicism since its logic was open to welcome another tradition while retaining their own.[10]

Christian history depicts numerous instances of a creative encounter with varied cultures at different stages of its evolution. The influence of West Asian cultures and customs is well documented in the Bible.[11] The history of Christianity in the early centuries is marked by the impact of Hellenistic, Roman, Germanic, and Celtic cultures. These cultural influences were taken for granted in the period of philosophical and doctrinal articulation of the Christian tradition with little attempt to critically examine the impact of such influences. The missionary movement from the sixteenth century did raise the issue of the gospel in relation to the newly discovered cultures of Asia, Africa, and the Americas. But unfortunately, the eagerness to "conquer" those peoples for Christ left little space for respecting or accepting the cultural richness of these people, as is evident in the case of Francis Xavier.

The Second Vatican Council, with a radical shift in orientation, acknowledged the plurality of cultures and spoke of entering "into communion with various cultural modes, to her enrichment and theirs

9 Schreiter, *Local Theologies*, 150.

10 Felix Wilfred explains inclusivism as a characteristic Asian trait in the understanding of a religious community. See his *Sunset in the East: Asian Challenges and Christian Involvement* (Madras: University of Madras, 1991), 235.

11 See Roland De Vaux, *Ancient Israel: Its Life and Institutions*, trans. John McHugh (London: Darton, Longman and Todd, 1986), part 4; Gerhard Von Rad, *The Problem of the Hexateuch and Other Essays* (London: SCM Press, 1984), chapter 14 and 15.

too."[12] It called for an extensive theological elaboration of the intimate relations between the church and cultures, and the mission of the church with regard to both society and culture (no.62). However, one cannot miss an overbearing attitude in these pronouncements. The same attitude is evident in the Apostolic Exhortation of Pope Paul VI (1975), *Evangelii Nuntiandi*, which called for "evangelisation of cultures," or in Pope John Paul II's call for the Church's dialogue with the cultures of our time.[13] Marginal people tend to look at notions like inculturation, cultural adaptation, and dialogue with cultures as coming from elsewhere and having little relevance to their daily concerns. The notion of inculturation assumes that the gospel (presumably shed of its Western cultural garb) comes face to face with a local culture like the neithal culture. The very attempt to take a local culture as an empty pot in terms of faith is anthropologically untenable. Therefore, it makes better sense to speak of "intercultural encounter and dialogue," to be fair to the people involved.

In fact, it is an anomaly to speak of inculturation in the context of coastal Catholicism. As inheritors to a dual religious system, the coastal community is already well *inculturated*. As subjects of their lived religion, they have been directing a process of dialogue between two cultural traditions and worldviews. This "intercultural dialogue" is not something to be started afresh; it has been going on quietly through many generations. Two illustrations from the coastal world would clarify it better.

The first refers to the label of devil worship that the missionaries used to refer to the traditional ritual practices of the fisherfolk. The missionaries were following the demonology of Christian Europe of the Middle Ages, which depicted demons as horrifying creatures responsible for every evil and misfortune. This demonology had been largely an amalgamation of patterns from various traditions in the surrounding cultures in its early development. Etymologically too, the use of terms like devil, demon, and Satan as interchangeable indicates a process of fusion of concepts.[14] But the labelling of coastal rituals as devil worship might not have

---

12 Vatican Council II, *Gaudium et Spes*, in *Vatican Council II: The Conciliar and Post-Conciliar Documents*, ed. Austin Flannery (Bombay: St. Paul Publications, 1988), 53 and 58.

13 For more details on this from the Indian context, see Kuncheria Pathil, ed., *Religious Pluralism: An Indian Christian Perspective* (Delhi: ISPCK, 1991); M. Amaladoss, *Becoming Indian: The Process of Inculturation* (Bangalore: Dharmaram Publications, 1992).

14 See *The New International Dictionary of New Testament Theology*, 1975–76 editions, s.vv. "Satan", "Demon"; *Harper's Bible Dictionary* (Bangalore: TPI, 1994); *Encyclopedia of Religion*, s.vv. "Demons—An Overview", "Satan." The Hebrew Bible used the term Śāṭān to mean "adversary" (Job chs.1–2; Zech 3:1–2). In the inter-testament period, the idea of Śāṭān as *the* adversary of God emerged; he

disturbed the fisherfolk for the simple reason that the devil (or demon) does not have such a repugnant meaning in the Indian religious firmament. Demons are not specifically evil beings, though not sufficiently divine to be regarded as gods. A total identification of demons with evil is certainly absent; the boundary line between gods and demons is thin and often shifting. The paradox of a good demon and an evil god is present in Indian demonology too.[15] Besides, the fisherfolk may read the biblical account of the devil in Genesis very differently; for, were they not angels originally, but punished for disobedience to become devils?[16] Thus devils are seen as divine beings of a different grade and having features of both angels and devils, thus at times benevolent and more powerful. Such reinterpretations give them confidence in facing the perils of daily life and make religious practice more satisfying.

The second refers to the veneration of the dead. The Christian creedal formula affirms belief in "the resurrection of the dead and the life of the world to come."[17] We discussed earlier the deep influence of the neithal culture on the funeral rituals and related beliefs of the fisherfolk. The hope of continued existence after death makes the fisherfolk feel at home with the Christian doctrine that the body will be reanimated. Resurrection is

came to be identified with the powers of evil in the universe, probably under the influence of Persian ideology of dualism. Gradually, the identification of Śāṭān with the fallen angel and "the serpent" of Genesis in chapter 3 followed (Rev 12:7–9; Luke 10:17). The Greek word *diabolos*, usually rendered Devil in English, was used in the Septuagint to translate the Hebrew term Śāṭān. But the Devil is also used to translate another Greek word *daimōn*, which originally referred to a minor deity or a guardian spirit; but it got fused with the Jewish *schedim* (vengeful spirit) and translated into Latin as the Demon to mean the embodiment of evil. Thus, Christian demonology is the outcome of varied conceptions, especially from Greek and Judaic traditions, fused together. Some N.T. texts also clearly associate Demons with paganism and heathen worship (1 Cor 10:20–21; Rev 9:20; Acts 26:18; 2 Cor 6:16, etc.).

15 See W.D. O'Flaherty, *The Origins of Evil in Hindu Mythology* (Berkeley: University of California Press, 1980); Louis Dumont, *A South Indian Sub-caste* (Delhi: Oxford University Press, 1986), 449. "Asuras" appear in Hindu mythologies as demonic creatures and the chief opponents of the gods, but the distinction in terms of good and evil is not absolute. "Those who always win are gods, but because their adversaries are never really eliminated, the kinship of the gods and the demons always resurfaces", writes Paul Ricoeur, *Encyclopedia of Religion*, s.v. "Evil".

16 Rev 12:9; Eph 2:1–2; Genesis 3. A similar conception is present in some Hindu texts too; according to *Amarakosa* 1.1.7, the Demon is called *pūrva-deva*, ie. "one who was formerly a god". See O'Flaherty, *The Origins of Evil*, 65.

17 This is the phrase used in the Niceno-Constantinopolitan Creed (381 CE), popularly known as the Nicene Creed. The version in the Apostles Creed reads, "the resurrection of the body and the life everlasting."

understood as an act of divine power in calling persons to a new form of living eternally in God's presence. It affirms that the glorious resurrection of the dead takes place when Christ appears in his majesty. At the heart of this doctrine is the belief that the crucified Jesus has been raised from the dead to become the Savior. 1 Cor 15:1–58 presents the argument logically.[18] But the intricacies of a bodily resurrection would certainly be baffling. For, if one continues to live after death, although in another mode of existence as *muni* or *peiy-piśāśu* or any other, resurrection becomes irrelevant. Moreover, the promised resurrection "on the last day" would appear too far off to become meaningful here and now. The fisherfolk may be trying to do away with the mythical garb of resurrection coming from another cultural tradition while affirming the Christian conception of "life everlasting."[19] They offer the Holy Mass for the dead at all occasions and perform elaborate rituals of *kuzhiōthal* on All Souls Day for the wellbeing of the dead who continue to live in one mode or another. Probably, the unity of the living and the dead that Christian doctrines affirm is better grasped by Neithal Catholicism.

These examples illustrate the ongoing encounter and dialogue between two religious traditions on the seashore. These do raise questions about Christian eschatology, demonology, and the doctrine of the bodily resurrection and point to their inadequacy in responding to the vital concerns of the coastal people.

Ambivalence of the Māta Image

Discussion on the cult of Cintāthira Māta in an earlier chapter unveiled two aspects that are central to Neithal Catholicism. First, it showed that the long tradition of the worship of a mother goddess or Śakti continues to survive and influence the community. Second, it indicated that Cintāthira Māta embodies the two contrasting features, viz., the benign and the terrible. The cult appears to be the product of their creative mind to deal with the apparent incongruity of superimposing a Mary image over their goddess heritage. The Māta has come to symbolize the community's identity in a powerful and unique way. This section tries to

18 The same idea is presented in Vatican II, *Lumen Gentium*, nos. 50–51.

19 The history of the Christian tradition shows that belief in life after death, especially belief in the resurrection, is a rather late addition to the body of Jewish beliefs. It is only in the book of Maccabees that explicit mention of it appears, probably necessitated by the mass killings of the Jewish people for their resistance to *paganism*. Moreover, the celebration of the All Souls Day was brought into the Roman rite only in the tenth century; in Celtic regions, some aspects of the pagan winter festival of *Samhain* were attached to this feast. See Victor Turner, *The Ritual Process* (Middlesex: Penguin books, 1972), 171.

explore how the traditional Catholic image of Mary gets reconciled to the goddess, and how the phenomenon raises fresh questions for the Church and for Christian theology.

In the Catholic tradition, Mary, mother of Jesus, is depicted as virgin, mother, queen, and bride. All these portray her as a benevolent, loving, and forgiving woman, and she remains the embodiment of all desirable feminine qualities. This is evident in the doctrinal history of the mystery of Mary that developed in close association with the christological doctrines. The Council of Ephesus (431) stated that Mary was not merely the mother of the man Jesus, but the mother of God himself (*Theotokos*), since God truly became a man through her. This divine motherhood has been the foundation for later doctrinal formulations on Mary. On a parallel stream, her virginity also came to be emphasized. Vatican II (1964) presents her as "an eminent and unequalled example of both motherhood and virginity."[20] The Apostolic Letter, *Mulieris Dignitatem*, written on the occasion of the Marian Year (1988) affirmed that "virginity and motherhood" coexist and wonderfully complement each other in Mary, although it would appear a paradox to the unbeliever.

The cult of Mary might have appeared in the Christian liturgical tradition only in the fourth century, but forms of Marian devotion existed earlier. This is not surprising given the Jewish heritage of strict monotheism devoid of gods and goddesses. But once Christianity got established in the Greco-Roman world, the influence of female deities from the surrounding cultures crept in. Historically, Mary exhibits continuity with the attributes of some West Asian goddesses like Isis, the patroness of sailors in Egypt, the Sumerian goddess Inanna (Queen of heaven and earth), and the Semitic goddess Ishtar, especially in combining in her the seemingly contradictory aspects of virginity and motherhood. The Cathedral of Our Lady in the historic city of Syracuse in Sicily was originally a temple dedicated to the Virgin Goddess Athena built in the fifth century BCE. In the seventh century CE the Byzantine Emperor turned Athena's temple into his Cathedral.[21]

Given this background, how does the picture of Cintāthira Māta fit into the portrayal of Mary in the Catholic tradition? The religious mind of the fisherfolk might have felt at home with the ambivalent nature of Mary as Virgin and Mother, for, the pre-existing Devī image was akin to such ambivalence. But the incongruity between the benign and the fierce

20 Vatican Council II, *Lumen Gentium*, 63.

21 See *New Catholic Encyclopaedia*, s.v. "Marian Feasts"; *The Encyclopaedia of Religion*, s.v. "Goddess Worship—an Overview"; Andrew M. Greeley, *The Mary Myth: On the Femininity of God* (New York: Cross Road Book, 1977), chapter 6.

would create difficulty since the fierce nature is so alien to the Mary of the dogmas. It is possible to see Cintāthira Māta as the creative answer of coastal Catholics to the apparent incongruity. The Māta thus becomes not merely a benevolent mother but also assumes the terrible features of a Devī. In other words, the Māta may be seen as depicting not only the benign nature but also the fierce features that are so characteristic of the pre-Christian Devī. Both the aspects seem to coexist in her.[22] The following steps may explain this process better:

a) The instance of fascination: The image of Mary resonates with the religious imagination of the fisherfolk due to her close resemblance to the Amma image that is so central to their tradition.
b) The instance of disenchantment: The image of the unidimensional Amma that depicted only the benign nature appears lifeless, and so not gratifying. They long for a "two-dimensional Amma" that reflects the very cosmic realism.
c) The instance of ascription: The missing elements in the Catholic Amma, especially the fierce features, are appropriated from the Indic/neithal tradition and ascribed to the Mary image giving it wholeness.

Thus, the Māta assumes the role of a protector goddess and even a mighty warrior in times of distress and crisis like that of a communal clash. The fisherfolk do not experience any serious conflict in combining the two contrasting features in the Māta. Still, the predominant figure of Cintāthira Māta is far from the fearsome figure of a Kāḷī; the overbearing presence of warrior saints like St. Michael and St. James, and the Catholic tendency to deny in Mary anything terrible may explain this. The cult of Cintāthira Māta can be seen as the product of a creative encounter and dialogue between two worldviews designed by the fisherfolk. In fact, a reverse process of inculturation may be at work here. Cintāthira Māta, for the fisherfolk, may mean a Christianized Devī rather than an inculturated Mary. This reverse process of inculturation has not attracted the attention of Christian theologians.

---

22 See Kalpana Ram, *Mukkuvar Women*, 63–65. Ram is partly right when she exposes the Christian splitting of the feminine into benign and malign forms depicting the Virgin Mary as the "embodiment of a purely benevolent idea of the maternal feminine." But she fails to see how the fisherfolk refashion Mary according to their own image in Cintāthira Māta. Similar phenomena of Marian devotion are present also in other parts of the world with a colonial past; examples are the image of *La Conquistadora* (Our Lady of the Conquest) in Santa Fe and that of Our Lady of Guadalupe in Mexico City.

The cult of Cintāthira Māta may also question the conventional portrayal of Mary as unidimensional. The one-sided accent on the benign, imposed on the religious cosmos of a people is anthropologically not sustainable. All reality, even sacred reality, comprises not only the good but also the not so good and the fierce, as Kadalamma reminds them day and night. The coastal reading of Amma may find some resonance with the gospel presentation of Mary in the Magnificat where she owns up a God who is not only benign but also fierce (Luke 1:46–55). Unfortunately, the later church traditions have underplayed this reading. Cintāthira Māta may open the door to a more authentic and balanced portrayal of Mary. The same logic would apply to the Christian conception of God in exclusive masculine categories. Neithal Catholicism illustrates the powerful human longing for the divine feminine within one's religious universe. No civilization in the world is said to have developed goddess worship as elaborately as in India. The Māta of the Indic/neithal tradition highlights the feminine aspect of God and the motherliness of the divine. Christian theology continues to struggle with the concept of masculine monotheism that it inherited from its Jewish roots. The fifth century Christians at Ephesus were preoccupied with the notion of "Mother of God," but what is to be recovered today is the very "Motherliness of God."

The Yearning for Warrior Divinities

The portrayal of Cintāthira Māta as the embodiment of both the benign and the terrible features indicates an inner longing in the coastal Catholics to access the wrathful dimension in the divine. This is further corroborated by pervasive veneration of saints with militant features in the coastal communities. More popular among them are St. Michael the Archangel, St. Sebastian the soldier-martyr, and St. James. The first two depict explicitly militant features. St. James is transformed into a military hero, in his local title as "Santiago who dispelled the enemy ranks" *(paṭai-mirattiya Santiyagappar)*. Militant features are attributed to St. Antony too who has no such background historically or scripturally. The terrible features of an otherwise benevolent St. Antony are portrayed in prayers and mantric formulae aimed at annihilation of evil forces. All these indicate the strong yearning of the fisherfolk for saints who depict fierce or warrior features.

The Indian theory of aesthetics (*rasa* theory) may shed better light on the attempt of the fisherfolk to incorporate the divergent features in their practice of Catholicism.[23] *Bhāva* is a core concept of this theory and is

23 It is true that the *rasa* theory has evolved in the sanskritic tradition in India, away from the neithal culture or any marginal context. However, this theory

widely used in Indian psychology, aesthetics and *bhakti* literature; literally, it means sentiment, disposition, or temperament, and may indicate the expression of a characteristic emotional state. It was Bharata (second century BCE to second century CE?) who systematized *bhāvas* and the related *rasas* in his treatise on dramatics *(Nātya Śāstra)*. *Bhāva* is what expresses a reaction, be it by bodily gesture or by words. Its end is *rasa* or the final state of relish or satisfaction. *Rasa* results from *bhāvas*.[24] He classified eight primary *bhāvas* that go with the eight *rasas*; they are love (*rati*), humor (*hāsya*), compassion (*karuṇā*), dread (*rudra*), valor (*vīra*), fright (*bhayānakā*), awe (*bībhatsā*), and wonder (*adbhuta*). Later theoreticians added a ninth one, viz. tranquility (*śānta*). Besides these primary sentiments, there are 33 (34) subordinate sentiments that serve to strengthen the prevailing sentiment.

Of the nine-fold categorization of primary sentiments, *śānta* (tranquility) seems to have acquired pervasive dominance in the Christian tradition, almost to the exclusion of most other *bhāvas* perhaps with the exception of *karuṇā* (compassion).[25] The fisherfolk, on the other hand, value other *bhāvas* like *rudra* (dread) and *vīra* (valor), which rarely find a place in the Catholic devotional landscape. *Bhāvas* like *adbhuta* (wonder) and *bhayānakā* (fright) have prominence in tantric rituals too. Some of the *mantras* used in healing rituals use the language of fire and sword aimed at the destruction of one's enemy. That the fisherfolk, with their hunting instincts, fighting with the ocean day and night, attribute these *bhāvas* to the Christian saints and to Virgin Mary points to their yearning for a religiosity that harmonizes with their daily life. It is also proof of their ability to adapt and re-present the universal Christian symbols and precepts to their life context and worldview. It could also mean a silent protest against the neglect of many basic *bhāvas* in Catholicism—in its sacred liturgy, its sacramental rituals, and devotional practices.

Some issues of theological concern surface here. The fisherfolk place before us the ingredients of a "theology of combat" as evolving from their very life of combat with the ocean, the wind, and the waves. Religion and theology cannot remain aloof from their daily life and concerns. Their love for Mother Ocean does not deter them from fighting her. Love and

presents one of the most comprehensive systematizations of human *bhāvas* that would make sense in any human situation. This fact would explain why the *rasa* theory is employed in the present analysis.

24 For a simplified exposition of the *rasa theory*, see Adya Rangacharya, *Introduction to Bharata's Nātya Śāstra* (Bombay: Popular Prakasan, 1966), chap.10.

25 This does not mean that Christianity is totally lacking in the expression of other *bhāvas*. The Old Testament clearly presents a wrathful and warlike God. See Samuel Rayan, *The Anger of God* (Bombay: BUILD, 1981).

fight coexist, both in life and in the practice of religion. There is a time to be tranquil and a time to be fierce. Many pastors got into trouble with the coastal people since they failed to grasp the basics of this theology of combat. Similarly, the possibility to perceive the Godhead in its multifarious *bhāvas* may be the Indian clue to the Trinitarian mystery.

The coastal approach to evil and suffering also opens up an interesting vista. The sharp line with which evil is contrasted with good, characteristic of dualistic thinking, seems to vanish on the seashore. Experience of evil and its ill effects is part of their daily life. The warrior saints and the mantric rituals are valuable assets in encountering evil. What is significant is that evil is accounted for within the sacred world, not outside of it. The Christian theological tendency to place God outside and above the realm of evil has its base in the Aristotelian-Scholastic philosophy. If evil is the opposite of good, and God is understood as all-good, then evil has to be outside the divine realm. This logic demands that evil is taken out of the existential context to a cosmic plane, and is attributed to Satan as evil personified.[26] But the fisherfolk experience evil at the existential level and try to explain it within their sacred world. The reality of evil is more complex than mere black and white, as the *rasa* theory illustrates. In fact, the coastal insight into demonology appears closer to the biblical viewpoint. Chapter one of Genesis presents God as struggling with evil (chaos) in the process of creation. Evil is recognized as a primal reality, and not necessarily as the opposite of good. Besides, analysis of the rituals reveals that the coastal people attribute the cause of evil to human agency, not to an abstract "evil one," nor to impersonal *karma*. The spirits have only an instrumental role, if any. The attribution to human agency brings a protective or remedial action within human reach. The fisherfolk, as subjects, thus refuse to submit to evil, and this may be the secret of their daring spirit and their resilience in adverse situations, whether in the ocean or on land.

An Experiment in Christian Tantrism

The widespread practice of mantric rituals (*mantravādam*) in coastal communities deserves probing and may help to explain the dynamics of a dual religious system. People generally accept *mantravādam* as powerful but know that it doesn't have the approval of the Christian community,

26 See Neuner J. and Dupuis J. ed., *The Christian Faith* (Bangalore: Theological Publications in India, 1992), 14. The Fourth Lateran General Council (1215) declared, "For, the devil and other demons were indeed created by God naturally good, but they became evil by their own doing. As for man, he sinned at the suggestion of the devil."

probably due to its condemnation by the official church.[27] Many people express guilt for utilizing mantric rituals. "Some of those who go for daily Mass and communion are the ones rushing to the *mantravādi* in the evening," some lament. It is illustrative of the incongruity of a double world in which they find themselves. An inner conflict does disturb many individuals and also the community as a whole. How do the fisherfolk deal with this dilemma is a pertinent question. One thing is obvious; many mantric rituals have been given a Christian form and content. The *Cakra* (wheel-seat) of St. Michael is a good example. Two imageries used in it are very biblical: the cross of Christ in its saving aspect and the sword of St. Michael in its fearsome aspect. The cross of Christ, in the Christian tradition, always stood for humble submission to the will of the Father. It evokes the benign aspect of the divinity. The sword of St. Michael, on the other hand, has reference to the Genesis story (Gen 3:24) and is the symbol of the wrath of God.

The preceding discussion establishes the deep influence of Tantric tradition in the everyday ritual life of the fisherfolk. Far from being two religious systems in conflict, the community tends to juxtapose them to make better sense out of a chaotic and risk-ridden world. The attempts at Christianizing the *cakras* and *mantras* are extremely valuable indicators of the way the coastal religious mind works. The deep emotive and cultural content of these for the people cannot be brushed aside. Is it Tantrism in Christian garb? Or is it vice versa? Is it possible to bring both Christian faith and Tantrism into harmony without serious conflict? Official Christian theology has so far paid little attention to these questions. But these are vital questions that emerge from a vibrant Neithal Catholicism that theology can no more ignore.[28]

In fact, Christian *mantravādam* is not a totally new concept. Ulloor Parameswara Iyer has a section on *Christiani mantravādam* in his book

---

27 See Scaria Zacharia, ed., *The Acts and Decrees of the Synod of Diamper, A.D. 1599* (Hosanna Mount: Indian Institute of Christian Studies, 1944), 202–212. The intolerant attitude of the missionaries to such practices of the Indian Christians is evident in these decrees. Lev 19:26 b and 19:31 unambiguously prohibit "practice of augury or witchcraft," and "turning to mediums or wizards."

28 See W. Teasdale, *Toward a Christian Vedanta: The Encounter of Hinduism and Christianity According to Bede Griffiths* (Bangalore: Asian Trading Corporation, 1987), 146–153. He discusses the efforts of Bede Griffiths to see the link between Tantrism and Christian practices in the context of inter-religious dialogue in India. He places Tantra as the opposite of the Vedanta, as a necessary requirement to balance the trans-worldly orientation of the latter. Tantra, according to him, is not something to be added on to the Christian tradition, rather, something that is already there, but only to be discovered. He makes a preliminary attempt to read the New Testament from a tantric point of view.

that discusses its practice among the Syrian Christians of Kerala, and quotes from Christian *mantras* and mentions practitioners like the legendary Kadamattathu Kathanar.[29] Even in the Western Christian tradition, the use of sacred monograms (IHS, INRI, XC, A–Ω, X–P, etc.) and palindromes[30] are indications of a similar practice of symbolic use of words. Early Christian art used many pagan motifs that were open for a Christian reinterpretation. The Greek word for fish (ΙΧΘΥΣ) is an acrostic form for "Jesus Christ Son of God Saviour," and it was a favorite Christian emblem especially of the Eucharist. Similar was the case with other motifs like the dove for peace and peacock for immortality.

Neithal Catholicism opens up the arena for a rethinking on our understanding of mantric rituals in the Christian tradition. Observation shows that most of the Christian practitioners of it use it in close association with the profession of healing. They practice one indigenous system of medicine or another, and many of them are local medical practitioners (*vaidyan*). Some authors indicate that exorcism and various mantric rituals were part of the treatment of physical ailments in the ancient Tamiḻakam before and even during the Sangam period. Sudhir Kakar discusses Tantrism with reference to the healing traditions of India.[31] Indic systems of medicine like Ayurveda and Siddha are considered as both a spiritual science and a medical discipline. The cosmic aspects of illness, ill fortune, and ill luck are highlighted by all Indian systems of medicine including Tantrism, thus making medical and "priestly" functions hardly separable. *Vaidyan* (literally, doctor) is usually both a medical man and a priestly (ritual) functionary.[32]

Two aspects of Tantrism seem to be of particular significance to Christian religion and theology. One is its emphasis on Śakti, the divine feminine, and the other is the affirmation of the human body. The Śakti aspect was discussed earlier in relation to the Cintāthira Māta, and so we

29 Ulloor Parameswara Iyer, *Kerala Sāhitya Charitram*, Vol. 3 (Thiruvananthapuram: Kerala University, 1972, 246). See Sebeena Raphi, *Chaviṭṭunāṭakam* (Kottayam: NBS, 1980), 65, for details of a *mantra* that was popular among the Syrian Christians of Kerala.

30 See J.C.J. Metford, *Dictionary of Christian Lore and Legends* (London: Thames and Hudson, 1983). *Palindrome* is a mysterious construction of words that can be read forward and backward. *SATOR* is the first line of a Latin palindrome found in Pompeii and in England; it is in the form of a square and is not yet clearly deciphered.

31 Sudhir Kakar, *Shamans, Mystics and Doctors—A Psychological Inquiry into India and its Healing Traditions* (New York: Alfred A. Knoff, 1982), 151–155.

32 See Aloysius Pieris, "Prophetic Humour and the Exposure of Demons", *Vidyajyoti Journal of Theological Reflection* 60 no. 5 (May 1996): 311–322. He insightfully reflects on a traditional Buddhist healing ritual.

now focus on the latter. In the Indic view, the human body is perceived as intimately connected with the cosmos and cosmic phenomena as made clear in all Indian systems of health and healing. According to *Caraka Samhita*, an ancient medical text, "The person is comparable to the cosmos... That is, the person is a minuscule image of the great cosmos. All the features (*bhāvas*) that are present in the cosmos are present in the person. All that are in the person are in the cosmos."[33] There is obvious continuity between the human body and the natural environment. This insight from Ayurveda underlies the Indian thinking in general.

The tantric tradition emphasizes the creative ordering of the powers of nature and of the body. In other words, the tantric tries to interiorize the cosmos through *cakras* and *yantras* aiming at transcending the duality of the inner and the outer. It is not an escape from the world here and now, but an affirmation of the world and it is this affirmation that makes Tantrism appealing to ordinary people in their daily life and struggles. The Christian image of the body, on the other hand, sharply differentiates it from the rest of the cosmos, especially the physical environment. The Christian idea speaks of the human person as created "in the image of God" (Gen1:27), implicitly underplaying the vital link with the rest of the created world. As a result, the human body plays only a minor role in Christian liturgy and related rituals. This may explain why the fisherfolk, especially the youth, look forward to the colorful procession at the parish festival to dress up and to celebrate the body as distinguishable members of pious associations (*sabhakal).*

Insights from *mantravādam* and Tantrism call for a conceptual reframing of the human body in Christian thinking with obvious implications for sacramental theology and spirituality. To accept the human body as the alter ego of the cosmos is to accept it as a primal sacrament. As a microcosm, it embodies the human-divine mystery. It is through the mediation of the body that human beings grow into this awareness, as the mystery of the incarnation, the mystery of the Word becoming Flesh, constantly reminds us. The incarnation invites human beings to celebrate the body as central to any act of worship. Themes like *body spirituality* and *vibration spirituality* call for more theological exploration. Any attempt to contain the tantric stream within the Christian contemplative tradition would be misguided.[34] Tantrism, with its accent on the microcosm and macrocosm, implies a clear affirmation of the body and the world, and cannot be correlated with the otherworldly emphasis of the

33 As quoted in Kakar, *Shamans, Mystics and Doctors*, 293.

34 Teasdale seems to be following this misguided direction; see Teasdale, *Towards a Christian Vedanta*, 170.

contemplative tradition. The liberation potential of Christian Tantrism is evident as it involves resistance to passive submission to one's fate, and an urge to tame the ocean—and oneself.

CHAPTER 8

THE STRUCTURE OF LIVED CATHOLICISM

The previous chapters have explained how the complex world of the rituals and beliefs of the fisherfolk weave together to provide a coherent universe of meaning as they live their religion in everyday life. To a large extent, the community has succeeded in reconciling the divergent elements coming from its neithal past as well as from its Catholic heritage. We saw that Neithal Catholicism has been the result of this process. The present chapter is an attempt to explore further how the fisherfolk organize their Neithal Catholicism in a way that would sustain them in daily life.

Every religious tradition tends to give rise to specific forms of organization, usually referred to as the process of institutionalization. It ensures continuity of the tradition and its ethos, besides strengthening the cohesion of the community. In the words of O'Dea,

> There are two chief kinds of religious organization found in human societies. In primitive and archaic societies, religion is a diffuse phenomenon; many forms of human association from the family to the workgroup have in some important respects a religious character. In these societies, religion is one aspect of the life of all social groups.... The specifically religious organizations are found generally in societies in which internal differentiation of function and consequent stratification have developed.[1]

World religions like Christianity and Islam usually emphasize the latter, namely the specifically religious organizations. In Catholicism, the organizational form is highly institutionalized and uniformly applied the world over, to a large extent. But this uniformity is mostly at the level of manifest religion, but religion as really lived may tell us something very different. At the lived or latent level, Neithal Catholicism operates mostly as a "diffuse" phenomenon, often indistinguishable from the people's

---

1 Thomas F. O'Dea and Janet O'Dea Aviad, *The Sociology of Religion* (New Jersey: Prentice Hall, 1983), 38.

sociocultural or occupational concerns. Many archaic societies manifest the same feature where material culture closely corresponds to the social or religious organization. In fact, the notion of a religious organization as a distinct category itself may appear anachronistic. It is the neithal culture, rather than any religion as such, that would better explain the organizational features that find expression at the religious realm.

The sense of solidarity, or the clan ethos, that characterizes the neithal heritage does have an overwhelming influence on the organizational realm of Neithal Catholicism. This is evident in the multitude of confraternities or pious associations (*sabhakal; sabha* is singular) in coastal parishes, and these blend the two religiocultural traditions in a unique way. It is interesting to see how these *sabhakal* act as cultural substitutes in coastal Catholicism.

Sabhakal as Cultural Substitutes

The Catholic fisherfolk along the coast take pride in the list of pious associations attached to their church. Vizhinjam provides us with a good picture of this phenomenon and is paradigmatic. Fourteen *sabhakal* were active and vibrant in the parish community as the data we gathered in 1995 indicate. All these bodies function with the approval of the church, and some are affiliated to diocesan federations. While pious associations or confraternities have always been part of Christian tradition universally, what makes the Vizhinjam scene significant is their proliferation having a deep impact on their religious life as well as other areas of their sociocultural life.

Each *sabha* has its own unique features both in its social identity and its ritual expression. The unique costume/uniform makes each group stand apart, and members are eager to display the full attire on designated days. They come alive on the feast day of the group's patron saint and on major feast days of the parish. For example, the first Friday of every month is a special day for members of Thiru Hrudaya Sabha; members stay away from fishing on that day and participate in Holy Mass and communion. Others also follow their example out of devotion and the seashore would look deserted. Some groups conduct their devotional practices in the church before or after Mass, or even while the Mass is in progress. It would appear that the group's ritual is more important to them than the official liturgy. The parish priest is the official patron of all *sabhakal*, but in practice, each group functions as an autonomous entity.

Sabhakal become more active during parish festivals and whenever a death occurs in the village. If the deceased is a member, then all the members of that *sabh*a in full costume visit the family and join in the

Chart: *Sabhakal* at Vizhinjam—An Overview (valid for Oct-1995)

No.	Name of Sabha	Members (approx.)	Nature	Patron Saint	Uniform
1	*Carmalamātha sabha*	36	men-adult	B.V. Mary	white cope, waistband, shawl, crown
2	*Jnānaprakasiar sabha*	75	men-adult	St. Aloysius	white cope, blue shawl, crown
3	*Thiru Hṛudaya sabha*	300	men & women	Sacred Heart	white dress, red ribbon, medal
4	*Mikēlmālaka sabha*	80	youth (m)	St. Michael	green shirt, white dhoti
5	*Geevarghese sabha*	80	youth (m)	St. George	white dress, red ribbon, medal
6	*Ouseph Pithau sabha*	30	youth (m)	St. Joseph	white dress, white ribbon, medal
7	*Antōṇiyar sabha*	100	youth (m)	St. Antony	brown cope, cincture, picture
8	*Almāya Carmalita sabha*	30	men	B.V. Mary	white dress, scapular
9	*Almāya Carmalita sabha*	30	women	B.V. Mary	white dress, scapular
10	*Vanakamātha sabha*	75	women	B.V. Mary	white dress, blue ribbon, crown
11	Vincent de Paul society	25	men & women	Vincent de Paul	white dress, scapular
12	Maria Goretti sabha	100	girls (-18)	M. Goretti	red blouse, white skirt, veil
13	Sodality	50	women & girls	B.V. Mary	white dress, green ribbon, medal
14	Legion of Mary	500	men & women	B.V. Mary	sky-blue dress or blouse/ shirt

funeral procession. Some groups accept invitations to accompany the funeral procession of any person on payment of a stipulated amount; it is both a prestige issue for the family and a source of income for the *sabhakal*. But this custom is slowly vanishing. *Compiria Sabhakal* have a prominent position in comparison with the other pious associations. *Compiria* or *compiri* is a term coming from the Portuguese Catholic tradition and is found being used widely not only in the coastal churches but also in other churches of both Latin and Oriental rites. Carmalamāta Sabha and Jñānaprakāsiar Sabha are *compiria sabhakal*. During parish festivals, members dress up in their attractive uniforms and accompany the procession, and the *compiria sabhakal* have the right to stay close to the statue of Cintāthira Māta. Further, the status position of a *sabha* reflects the status position of its members in the coastal community. Almāya Carmalita Sabha and Vincent de Paul Society are identified with rich people, while others like Antoniār Sabha are treated as the monopoly of certain lineage families. Some are dominated by the youth and are of recent origin. Membership is confined mostly to the youth who are fishermen and they are poorly educated. These *sabhakal* often appear like social clubs as the members have little access to other sociocultural or political groupings. We may argue that the religious appeal draws these youth to the *sabha*, but the fact is that the *sabha* is mostly a response to their unmet need to belong.

Traditionally, the *compiria sabhakal* had officials like Pirusenti (president), Procudor (procurator), Iscruman (scribe or secretary), and Chemmador (thurifer). The functions may vary from place to place. Pirusenti, in many places today, is the person who is elected to take charge of the parish feast for a year. But traditionally, the Pirusenti is the leader of a *sabha* and is elected for a year. He has no responsibility for the parish feast. A ritual celebration called *Pirusentikalyāṇam* is the ritual of the Pirusenti assuming office. The *sabha* members solemnly escort the designated person who wears a jewel-studded crown and a decorated scepter, with the accompaniment of a bell, band set, and crucifix. The priest celebrant crowns him and hands over the scepter during the Holy Eucharist. It is the right of the outgoing Pirusenti to host a dinner. The ritual has other embellishments of a parish feast, like hoisting the flag a few days in advance and the accompanying ceremonies. Pirusenti is are entitled to use the honorific title with their name all their life.

There are indications to believe that the office of Pirusenti meant more than being the leader of one *sabha*. It is likely that this office had in earlier days close links with that of Pradhāni (village chief). Śesadima Pirusenti D'Costa is still remembered as village chief and community leader. He was the grand uncle of Renkayyan, the last Pradhāni who is still alive. It is probable that in the past, the office of Pirusenti had appropriated many roles and functions of a community leader, which in course of time got reduced to that of a *sabha*-leader. Deeper historical research is required to get better clarity on this critical area of community leadership.

The roots of these *sabhakal* can easily be traced back to the confraternities of the medieval European Catholicism. Confraternity is a term used usually to designate societies of lay people, canonically set up under ecclesiastical direction, for the spiritual welfare of its members. General or particular works of apostleship also are often associated with them. A confraternity usually has its origin in, and association with, devotion to a particular mystery of religion, a saint, or religious observance. Members wore capes of a different color to distinguish themselves. Many confraternities were promoted or directed by religious orders or congregations. Most of the older confraternities came into being during the later Middle Ages, but the majority of the present Marian confraternities date from the Catholic revival after the Protestant Reformation.[2]

Confraternities played an important role in the Christian missionary expansion in India and elsewhere. When Francis Xavier landed in Goa in 1542, the Confraternity of Mercy (*Santa Casa de Misericordia*) for the Portuguese devotees was already in existence, and it was engaging in relief for the poor in Goa.[3] Friar Diago de Borba developed an institution for training young boys who were later banded together into the Confraternity of the Holy Faith. This body played a key role in the organized conversion of the natives in Goa.[4] Its power and influence increased as the Portuguese government took over the property of temples they destroyed and set aside the income from it for the upkeep of confraternities and churches. At times, the duty to maintain law and order had also been entrusted to this confraternity. *Confrarias* were set up often along caste lines, thus reflecting the very caste composition of the Goan society. In the Confraternity de Santissimo de Nossa Senhora de Socorro, started by

2 See Donald Attwater, *A Dictionary of Mary* (London: Longmans, 1957). Rosary Confraternities were started in 1470 at Douay in Flanders and were promoted by the Dominicans. The Scapular Confraternities arose in the sixteenth century under the patronage of the Carmelite friars. The Sodality of the Bl. Virgin Mary was founded in 1563 at the College of the Society of Jesus, Rome, by Fr. John Leunis, a Belgian Jesuit, mainly as a gathering of pupils; soon it spread to all Jesuit educational centers and parishes everywhere, and became the model for other confraternities that arose later.

3 George M. Morais, *A History of Christianity in India: From Early Times to St. Francis Xavier, A.D. 52-1542* (Bombay: Manaktalas, 1964), 231.

4 R. de Souza, *Goa to Me*, chapter 2.

the Jesuits in Goa, the *Gauncars* (the landowning and ruling caste) alone were given membership; they wore a red cape, and they alone had the right to carry the cross on Good Friday.[5]

Along the fishery coast, the Confraternity of Charity was the earliest, started by Fr. Henriques in 1575 at Punnaikayal. Later, the Sodality of Our Lady was formed in Tuticorin in 1648 and gradually spread to other fishing villages too. The growth of a culture of *sabhakal* along the Thiruvananthapuram coast must be seen against this background. What originally started as an instrument of religious renewal later became a tool for conversion, and then a symbol of caste status in the religious hierarchy, and finally, a form of cultural association or club. Despite these transformations, the phenomenon of *sabhakal* presents a good example of a harmonious synthesis of Iberian Catholic confraternities and the Neithal clan ethos molded in religious garb.

The *sabhakal* in the coastal parishes of Thiruvananthapuram flourished in a way never seen elsewhere in India. Two reasons could be attributed to this. The first is social; the vacuum created by the elimination of pre-Christian clannish affiliations offered an ideal environment for them to take root. The coastal Catholics found the *sabhakal* resonating with their need to express their subcultural identity within a church framework that remained mostly cerebral. The second is cultural; pre-Christian art forms and cultural expressions were banned by the missionaries as pagan and devilish and no proper substitutes were offered.[6] The scope for performance that the *sabhakal* offer with their attractive costumes caught the imagination of people and filled the void. The second reason becomes more vivid in comparison with the coastal Christians in other parts of Kerala where effective Christian substitutes came to their rescue.[7] The Catholic fisherfolk in Alappuzha and Kochi have a similar religious background, and their Catholic history goes back to the same Portuguese period. Though pre-Christian art forms were done away with, the void was soon filled by introducing *chaviṭṭunāṭakam* and *parichamuṭṭukaḷi* that were given Christian content and flavor. *Chaviṭṭunāṭakam* was a Christian form of dance drama promoted by the Portuguese missionaries after the Synod of Diamper *(Udayamperur)* in 1599, as a technique to dissociate the coastal Catholics from "pagan"

5 Rowena Robinson, "The Cross: Contestation and Transformation of a Religious Symbol in Southern Goa", *Economic and Political Weekly* 29, no. 3 (January 15, 1994): 97.

6 *Mudiyēṭṭu*, for example, was a popular form of folk art in the villages of the Thiruvananthapuram coast and continues to be performed by Hindu villagers.

7 Sebeena Raphi, *Chaviṭṭunāṭakam* (Malalyalam) (Kottayam: National Book Stall, 1980).

rituals and practices like *kūthu, kūdiyāṭṭam, mantravādam, chāthanseva,* and animal sacrifice. *Chaviṭṭunāṭakam*, thus, was the product of a beautiful synthesis of the martial art forms of Kerala, the classical Indian dance movements, and the Greek convention of drama. A single performance would last fifteen nights consecutively; the time and expenses involved in such a task may be one reason why this art form has lost its popularity later. No similar attempts seem to have been made among the fisherfolk of the Thiruvananthapuram coast. Do *sabhakal* act as an escape mechanism that hinders a cultural resurgence of the coastal Catholics? This question needs probing.

Religious Control, Pradhāni, and Community Leadership

The phenomenon of the *sabhakal* as an interface between the neithal and the Catholic traditions must be situated within the cultural matrix of the community. An insightful area is that of religious leadership and related control mechanisms that operate in the Catholic community. Some concrete instances from the parish of Vizhinjam may illustrate the point.

After the communal riots of 1995, the church council issued a directive to the following effect:

i. Except for June, July, and August, fishermen from other villages should not come to Vizhinjam for fishing. (This was because the riot was said to have sparked off with a quarrel involving fishermen from elsewhere).

ii. No Catholic at Vizhinjam may enter into trade transactions with the Muslims of the area. Fish caught may not be sold to them, nor ice blocks bought from them.

These were announced at Sunday Mass. People adhered to these norms without a murmur. One fisherman who came from another village for fishing was beaten up. Two Christian petty traders who tried to sell fish to Muslim merchants through the back door were banned for two weeks. These measures were not unique to a riot situation but part of the ordinary system of controls in the coastal community. Oral history and the available church documents corroborate this. Some decisions of the parish committee during 1982–1984 can be illustrative:[8]

22.2.82: T.R. is fined Rs. 101 for misbehavior.

27.2.82: If anybody sells liquor to outsiders after 6 pm, he will be fined Rs. 101.

[8] Actual names of persons have been avoided.

27.2.82:	If anybody brings liquor from outside the village and sells it, he will be fined Rs. 501. The informer will be rewarded Rs. 101.
21.3.82:	Sale of liquor within the parish area is banned.
21.3.82:	If anybody initiates a quarrel, he will be fined Rs. 501. If the person happens to be a member of the parish committee, the fine will be Rs. 1001.
11.4.82:	T.A.C. is fined Rs.101 as punishment for riotous behavior after drinking.
27.5.82:	D.J., being a member of the parish committee, is fined Rs.1001 for quarrel after drinking.
22.6.82:	R.M. is fined Rs. 501 for selling liquor; a reward of Rs.101 was given to those who caught her.
25.7.82:	P.D. is fined Rs. 501 in liquor case.
20.11.82:	S.D., a member of the committee, failed to pay the fine of Rs.1001; so he is expelled from the committee.
23.6.83:	Catching fish using craft brought from elsewhere, or selling that fish in the area is forbidden for three months; fine for violation is Rs.1001. Informants will be rewarded Rs.101.
18.2.84:	Decided to pull down an unauthorized house that was put up near the seashore.
26.3.84:	Decided to collect sales tax (*kuthaka*) at the rate of 5 paise per rupee. (This was later abandoned due to resistance from the people).

The extract reads more like a judicial document or a police case diary. It indicates the overarching control of the church and the parish committee over not only religious matters but also over social customs and occupational practices including fishing. This gives us an idea of the powerful custom of ostracism or "community sanction" *(ūruvilakku)* still prevalent in some villages. In earlier days, it meant expulsion from the village territory (*ūru*). Other punitive measures included tying the offender to the flag post of the church and whipping, parading in public with hair shaved off, or with coconut shells or bone pieces tied around the neck. Such customs were gradually replaced by simpler punishments like kneeling before a Sunday congregation holding a cross. This too has been disappearing in recent times.

Published reports indicate the existence of similar practices in other coastal villages in earlier days. The *Latin Christian* of April 25, 1935 reported that a man at Valiathura was fined and asked to hold the cross during Sunday mass for violating a woman. The present-day forms of community sanction include a ban on work, fining, and cancellation of membership in church bodies. An offender is denied the sacrament of marriage in the parish church to avoid a scandal, though he is permitted to approach another church for the same. *Ūruvilakku*, being such a powerful instrument of community control, illustrates the firm hold of religion on the daily life of the fisherfolk. The sacred aura of these sanctions has helped the Catholic fisherfolk to maintain its identity and cohesion in an alien milieu over centuries by affirming its inherited wisdom and moral code in spite of the ill effects of these sanctions.

It is the realm of economic control that plays a vital role in support of the system of religious controls and sanctions. Two major means of economic control that are closely related to fishing are sales tax (ST) and donation. Sales tax is a sort of compulsory church tax on all trading operations connected with fish in the village and is popularly called ST or sales tax. The same is also known by other names such as *kuthaka (*monopoly right*)* or *kūṛu* (share). The rate varies from village to village. At Vizhinjam the rate in the mid-nineties was one percent (one paisa per rupee), and hence its popular name, "one-paisa collection." It is the responsibility of the buyer to pay the tax as it is collected at the sale stage. Besides, ST is also applicable to the sale of fishing craft and gear if the buyer is from outside the village. So, this is binding on the private boat-building yard that sells boats made of plywood and fibreglass to outsiders. The collection-right is auctioned annually by the parish committee and daily collections are made by collection agents. The auctioned amount at Vizhinjam in 1995–1996 was Rs. 11,52,000, a substantial amount of income for the church. The changing scenario has forced churches to do away with the system of sales tax, and it has been replaced by a system of voluntary donation in most coastal villages. The fisherfolk willingly keep aside a share of their catch for the upkeep of the church, and the amount thus paid is read out every Sunday after Mass. This system doesn't require a collection agent unlike in the case of ST. The crew of every boat entrust their "sacred share" *(nerchapirivu)* to the boat owner to be handed over to the church every week.

Traditionally, the tax was compulsorily being collected from coastal villages at the rate of five to ten percent. The transformation of *kuthaka* into "donation" at Vizhinjam has its own history. Until 1968, the *kuthaka* system continued at the rate of five percent, and at times even ten percent. But elders say that there had always been problems with *kuthaka*, either in terms of default by the auctioneer or being caught up in legal tussles. With the communal clashes of 1968, the general thinking began to change, and many began to feel that *kuthaka* was a kind of slavery to be rid off. Fr. Peter Borgia, the vicar at Vizhinjam from 1972, articulated the new thinking: "Let us do away with *kuthaka*; let each one contribute as he wishes." Thus, the new system of "donation" came to be accepted and is in existence even today. Records show that attempts were again made to collect ST at the rate of one percent in order to ensure a steady income for the church. Even in 1984, the church committee made a futile attempt to collect *kuthaka* at the rate of five percent.

The Catholic clergy in Travancore used to receive stipends from the Goa government in the earlier period of Portuguese presence. The royal treasury at Goa paid a subsistence allowance of Rs.750 to the ecclesiastical governor of Cochin and the twenty-eight vicars on the coast of Travancore and Coromandel received Rs.108 each per annum, and fifteen missionaries in Ceylon also received a similar salary.[9] However, these stipends were not sufficient to support the churches, and so other sources of revenue had to be found. Fishermen "were taxed with what was called *tithes* on the fishing nets, i.e. one-tenth of the fish caught in every net was given over to the church and the sale proceeds realized... . And this has become a national custom with them, and prevails to this day in all the coasting villages."[10] In Goa too, the Portuguese government used to extract *donativos* (euphemism for compulsory payments) from the natives for maintaining church institutions.[11]

History shows that in earlier days, *kuthaka* was used as a tool of religious control and exploitation of the fisherfolk. Old issues of the newspaper, *The Latin Christian*, carried a series of articles on the *kuthaka* system way back in 1935, and concluded: "The main reason for the troubles in the Portuguese missions of south Travancore is the *Matsyakuthaka*."[12] Whatever be the term used or the way it is extracted, the fisherfolk felt an inner compulsion to set apart a share for the church. "We will get a good catch only if we make a generous contribution to the church," seems to be a general conviction. The words of Markos, an elderly fisherman, are sharp: "Take away the money meant for the church, and you will be hit by a thunderbolt."The owner of a boat will be taken to task by the workers if he fails to pay the share of the church on time. "Let the defaulter be cursed, and let his craft and gear go down"—such words indicate the prime importance of church's share for the safety of fishermen. Some

9 See C.M. Agur, *Church History of Travancore* (New Delhi: Asian Educational Services, 1990), 277.

10 Agur, *Church History*, 277.

11 R. de Souza, *Goa to Me*, 73.

12 *The Latin Christian* (May 16, 1935), 1.

observe that voluntary contributions to the church at Vizhinjam shot up rapidly after the communal riots in 1995.[13]

The fisherfolk seem to equate *kuthaka* with the fifth commandment of the church. The offering of a tenth of all produce to Yahweh was a revered Old Testament custom (Deut 14:22–29; Lev 27:30; Neh 10:37 NRSV). The belief that the "impurity" of a substance will be eliminated only if a tenth of it is removed is also a strong New Testament motive (Matt 23:23; Luke 11:42). These biblical themes, together with the traditional attitude to the divine, have contributed significantly to the preservation of ecclesiastical control over the occupation and social life of the coastal Catholics. The vicar and the parish council have gained absolute control over the affairs of the fisherfolk. The vicar who is appointed by the bishop has a primordial role and authority in every matter pertaining to the community. His word is final and the Sunday pulpit provides him with a platform to announce decisions and directives with a sacred halo demanding absolute submission. The clearest example was the ban on the production and sale of arrack at Vizhinjam in 1993. The parish priest of that time has this to say:

> I do not know how it happened. I spoke out openly in the church about the evil of liquor and how it is destroying many of our families. I gave a call to put an end to this evil in our parish. The impact of it was immediate. Soon after the Mass, people rushed to the shops and houses selling liquor and began to smash the pots and jars in violent fury. Some men of the liquor lobby tried to resist but in vain.[14]

After two decades, the situation remains not very different as the experience of the vicar at Poonthura at the time of Cyclone Ockhi in 2017 illustrates.

> It was the second day after Cyclone Ockhi hit the coast. Nearly 50 fishermen from Poonthura were yet to return from fishing. The anxious villagers crowded around the church for a direction. Should we go in search of the missing men or should we not? The craving to go in search of the dear ones was compelling. But the officials had warned against venturing into the sea that was still in fury. All looked up to the vicar for the final word. I was in a dilemma. After a brief prayer in the church, I came out

[13] The 1995 communal riots involving Catholic fishermen and Muslim fishermen at Vizhinjam occurred first on May 14, followed by a second outburst on July 10. Six people, two Muslims, and four Catholics got killed, and over 200 houses and 100 peeling sheds were set on fire. Vizhinjam has a history of repeated communal riots; riots took place in 1959, 1968, 1972, and 1982.

[14] Personal interview with Fr. D. Stephen, vicar of Vizhinjam church, in September 1996.

> and announced: no one shall venture into the sea. There was total silence of acceptance.[15]

It is not surprising that people in many a village refer to their vicar as a little king (*kuṭṭirājā*).[16] The complex world of the control mechanisms and leadership structures in the coastal communities is an area that deserves closer scrutiny. Presently, we have to rely mostly on scanty historical records and comparative observations. The control structure implanted by Francis Xavier and his companions are well documented. The pattern, systematically developed among the Paravas of the Coromandel Coast, had been mostly transferred to the Travancore coast. Xavier appointed accountants (*kaṇakkapiḷḷas*) in the Travancore villages in order to support the work of the priests.[17] Today, the accountant functions as sacristan and custodian of sacramental registers also. Another office is that of *Meliñchi,* perhaps a corrupt form of *meirinho* that Xavier refers to in his letters and instructions.[18] *Meirinho* is the short form of *meirinho do monte* in Portuguese, meaning a warden or Portuguese law officer.[19] *Meliñchi*'s work, today, is that of a church attendant and consists of ringing the bell, informing the village if there is a death or a special Mass (going around ringing a bell with the announcement), and running errands for the vicar. He is maintained by the village on his customary share of fish. The *mūsk* was the solo singer in the church for liturgical ceremonies until recently when he was replaced by the church choir that sang in the local language. Francis Xavier speaks of other offices like *modom* (overseer of ecclesiastical details), *ubadesiār* (sacristan), and *vathiar* (teacher), but these offices do not exist in the churches along Thiruvananthapuram coast.

An important office having no reference in the missionary records is that of the Pradhāni (at times called *pramāṇis*). The actual executive powers were in the hands of this elite group of men who came from a few prominent families until they were replaced by the parish council. When the Basic Christian Communities (BCCs) were set up to decentralize power, the office of Pradhāni became defunct. Renkayyan Gera,

---

15 Personal interview with Fr. Justin Judin on September 20, 2018.

16 Kalpana Ram, *Mukkuvar Women*, 34. This stands in clear contrast with the phenomenon in Tuticorin where the JātiTalaivan, not the priest, was being referred to as the Little King. See Susan Bayly, *Saints, Goddesses and Kings: Muslims and Christians in South Indian Society, 1700–1900* (Cambridge: Cambridge University Press, 1989), 347.

17 Schurhammer, *Francis Xavier,* 474.

18 *Letters and Instructions of Francis Xavier,* trans. M. Joseph Costelloe (Anand: Gujarat Sahitya Prakash, 1993), 196, 356.

19 *Letters and Instructions of Francis Xavier*, 106.

the last Pradhāni who is alive, shared that the Pradhāni was the all-powerful leader *(sarvādhikāri)* of the coastal community at Vizhinjam. He recalls: "The priest had no power; he simply had to do what we asked him to do. We nominated even the committee members." The priest's duty was limited to saying Mass and spending money as dictated by the Pradhānis. The authority of the priest was only nominal. The Pradhānis decided every matter related to the community, whether buying property or putting up a church or settling disputes or fighting the Muslims. They decided on cases of ostracism and community sanctions (*ūruvilakku*). We may assume that the coastal community had a past with a leadership that existed relatively free from the clerical controls.[20]

The leadership structure in the Catholic community on the Coromandel coast will give us more insights. The Jāti Talaivan was the undisputed leader of the Parava Christians, and he presided over not only the religious and social affairs but also the political and commercial affairs of the community. In comparison, the Pradhāni appears to be a mere shadow of the vicar, the all-powerful religious leader. The legitimacy of his power came from his affinity to the priest and the religious structures. There is hardly any indication to prove his influence on sociocultural or political matters. Today, as Pradhānis have also become extinct, the office of the vicar gained dominance. The following observations with reference to Vizhinjam are illustrative, and are equally valid for other villages:

a) During the communal riots of 1995, the Christians were rushing to the vicar for leadership to fight the Muslim attackers. Some complained that they lost many lives just because the vicar (and the parish council) failed to supply bombs. Negotiation with government officials was done solely by the vicar and the parish council.

b) A perusal through the parish records shows that deliberations mostly referred neither to spiritual matters nor to church affairs. They read more like a police case-file, giving details of disciplinary action against persons on complaints about violation of fishing norms, misbehavior, or quarrels. Most of these should have been dealt with in the fisheries office or village office or police station.

c) The disproportionately thick "file of petitions" in the parish office is proof of the pivotal role of the vicar and the council in the

---

20 See J. Murickan, *Religion and Power Structure in Rural India* (Jaipur: Rawat Publications, 1991), 54–55. It gives a good description of the institution of Pradhāni as it exists at Poovar, a nearby coastal village.

> daily affairs of the fisherfolk. The file for 1995 (from January to August) alone has over 150 petitions addressed to the vicar. "The parish committee has become a *parāthi* (petitions) committee," laments a committee member.

The overall leadership of the coastal community, thus, is retained by the vicar. Besides religious and liturgical leadership, he has also become the undisputed leader of the community in sociocultural and political matters. Even government officials consult him first in any matter related to the coastal context as was demonstrated during the Kerala floods in August 2018. The Archdiocese of Thiruvananthapuram mobilized an army of fishermen with their boats who rushed to the flooded areas and saved thousands of lives, all at the request of the District Collector. A critical reading would tell us that the religious leadership, in the course of its Catholic history, has gobbled up the sociopolitical and cultural leadership of the fisherfolk. Mass conversion to Catholicism also resulted in the elimination of the existing structures of community leadership perceiving them as a possible threat to Christian hegemony. The power struggle between the two strands of leadership might have taken a long time involving many generations. In a reversal of history, the vicar today assumes the role of a "tribal chieftain," quite oblivious to the fact that this can be suicidal to the community in the changed political climate of the country. This collapse of community leadership is haunting many a thinking person in the fishing community today.

The Goan clergy was serving the coastal community of Thiruvananthapuram after the departure of the Jesuits from the scene. Goa on the west coast of India had become a major Portuguese center politically, commercially, and militarily. The Portuguese captured Goa in 1510 and made it their administrative, commercial, and ecclesiastical headquarters in India by 1530. Later the Court of Inquisition too was established in Goa. This intermixing of commerce and religion was further strengthened by the feudal worldview of Portuguese Catholicism. Rowena Robinson writes: "In the feudal worldview, religion (Catholicism) was the dominant force; politics, kinship, and economics all found their place in relation to it."[21] This background explains partly the mindset of the missionaries, and later of the Goan clergy, who held control over the fisherfolk of the Travancore coasts. The vicar had been conferred the triple powers over the caste, the village, and the church, if a declaration by a Goan priest at

[21] Robinson, "The Cross," 94.

Manapad is proof enough.[22] This was facilitated by the *Padroado system* that was in existence until recent times.[23]

The situation among the Paravas of the Coromandel was slightly different. The Portuguese retained the office and status of the Paṭṭankātins, the Adappans, and the Jāti Talaivan (caste headman) as key to maintaining the social structure and cohesion of the community. As a result, the Parava caste institutions were not eliminated, but rather got strengthened through their association with the Portuguese, and later the Dutch and the English. The same caste institutions and status system were carried over into the church context. Although there had been a long history of conflict over this state of affairs, the community leadership maintained its relative autonomy vis-à-vis the religious leadership. True, the Paravas had the advantage of a strong tradition of a system of status and authority even before their conversion. Among the fisherfolk of the Travancore coast, in contrast, the Portuguese had little commercial or political interest, and the missionaries were left to themselves. Their initial interest in Vizhinjam was only to use it as an outlet for shipping their pepper from Travancore. Having little economic or political base as a community, the fisherfolk easily came to be subdued by the religious hegemony of missionary Catholicism. The same state of affairs continues with little change even today.

Two Models of Religious Leadership in Conflict

An analysis of the process of encounter between the two religious traditions on the seashore makes it amply clear that the Catholic tradition contributed to the structural identity of the community while the meaning system remained deeply rooted in the Indic/neithal religious heritage. A process of encounter and dialogue has been taking place at the latent level giving new expressions to Neithal Catholicism. This process

22 Bayly, *Saints, Goddesses and Kings*, 376.

23 See *New Catholic Encyclopedia*, s.v. "Padroado of Portugal". *Padroado* or patronage was a form of benefice or reward granted by the Holy See to Portugal (and also to Spain) in return for their active role in the propagation of Catholic faith in the newly discovered territories during the colonial period. The benefice included the right of Portugal to present candidates for a bishopric or other ecclesiastical offices in these territories. This privilege was first granted in 1481 by the papal bull *Aeterni Regis*; in 1497, this was extended to the territories in India by Alexander VI through his bull *Ineffabilis*. It continued till 1950 when Portugal renounced the *Padroado* in Indian territory, though the final settlement came only in 1953. The coastal villages of *Trivandrum Mission* extending from Pallithura to Thengapattanam remained under the *Padroado* diocese of Cochin until 1955 when this area was annexed to the diocese of Trivandrum that was erected in 1937.

does not seem to have left any shattering impact on the community at the faith level. But it created difficulties for understanding of leadership in the community, particularly its religious leadership.

The mass dimension of the conversion process led to a total displacement of the existing structures of community leadership by a centralized system of religious control introduced by the missionaries. The office of the Pradhāni (village leader) and the institution of village assembly (*ūrukūṭṭam*) were either eliminated or made subordinate to the all-important office of the vicar. This has led to a crisis of leadership in the community. The priest came to assume not only religious leadership; the sociopolitical as well as the juridical leadership functions were also accrued to the priest. This centralization process eventually resulted in the suppression or retardation of various leadership functions in the coastal community. The heart of the crisis can be seen as a conflict of perceptions regarding religious leadership—between a centralized and hierarchical model, on the one hand, and a "diffused model," on the other.

The office of the vicar as the administrative head of a constituency called the parish has its roots in the feudal background of medieval Europe. From the fourth century, the local organization of the Western church followed the civil model of the Roman Empire. With the conversion of the Germanic tribes, the structures of the church came under the influence of the Germanic royal model in which the princes represented both the earthly and the sacred powers. The concept of the Roman Curia (Papal Court) thus evolved in the twelfth and the thirteenth centuries. The church was seen as the sole representative of every aspect of people's daily life, legitimized by the doctrine of the "Two Swords," the temporal and the spiritual. This state of affairs continued until the rise of independent nation states. This was basically the prevalent model at the time of the missionary enterprises worldwide in the sixteenth century. It was the same model that was implanted by the Iberian Catholicism among the fisherfolk of the Kerala coast. The past few centuries of its existence on the Indian shores could hardly transmute this model.

The centralized hierarchical model as described above must be examined in relation to the nonvedic model of religious leadership that existed in India. The absence of a regular priesthood characterized the religion of ancient Dravidians; in other words, religious functions remained "diffused" in the community, having no hereditary or professional priests.[24] There was no priestly class set apart, unlike in Vedic Brahmanism.

---

24 See P.G. Reddy and A. Chellaperumal, "Religion Among the Irulas of Coimbatore District, Tamilnadu", ed. Sudarsan et al., *Religion and Society in South India* (Delhi: B.R. Publishing, 1987), 76–86. Ethnographic studies highlight the

According to the Vedic tradition, priestly functions appear to have remained in the hands of a special class of hereditary priests who came to be known under the generic title of "Brahman." The "profession" was based on sacred learning and knowledge of the tradition. Among the different categories of priests who assisted at sacrifices, the *purohit* was the domestic priest at the service of kings and nobles.[25] The texts of the Brāhmaṇas show the *brahmans* emerging as a definite class in the order of caste hierarchy (*varṇa system*). As the Vedic sacrifice was replaced by temple worship in the later period, *brahmans* became temple priests. The Vedic model gradually penetrated the non-Vedic religious traditions in the course of history.[26]

The deep influence of this diffused model is evident in the coastal community even today. The so-called priestly functions are found dispersed among many functionaries. *Pirusenti* leads the affairs of the *sabha*, *chelāḷi* supervises the fishing operations, *vaidyan* or *mantravādi* attends to the healing needs and the occupational rituals, and *nāvier* serves at funeral rituals. These functionaries, at other times, are engaged in ordinary occupations like fishing or trading. Missionary documents show clearly that the fisherfolk had been resisting the priest-centric Catholicism time and again. It is probable that the model of religious leadership, and of the priesthood, that was inherited through the missionaries remained alien to the neithal religious sensibility, thus resulting in an ongoing conflict.

The coastal perception of the Catholic priest, at the same time, has an element of ambivalence. The priest is always revered and highly regarded in the community in a way rarely seen elsewhere. How can we understand this reverence against the perception of this model of priestly leadership as an alien? The ambivalence may be explained in terms of the distinction made earlier between identity structures and meaning frame. The office of the vicar is integral to the manifest form of religion, and hence a key symbol of its "identity structure." The alienation is felt at the meaning level, and so finds expression in the diffused model of religious leadership at the latent level. The conflict resulting from these divergent perceptions of leadership functions is not felt acutely, probably because the two operate at two distinct levels. The vicar finds himself largely excluded from the realm of latent religiosity. The wide

diffused nature of priestly functions in south Indian communities. The *Iruḷas* of Kerala–Tamil Nadu border on the Western Ghats, in spite of Vedic Hindu influences, still retain at least four categories of the priesthood, viz. *pattakaran, mannukkaran, pattali*, and *guruvan*.

25 The Ṛgveda (2.1.2) refers to different categories of priests who assist at sacrifices, viz. *hotṛ, potṛ, nestṛ, agnidh, prasastṛ, advaryu* and *brahman*.

26 See *Encyclopedia of Religion and Ethics*, s.v. "Priest, Priesthood" (Hindu).

gap that exists between the two raises questions about the adequacy of the religious priesthood in a marginal community and its congruence with the heritage of non-Vedic priesthood. Indian Christian ecclesiology may not remain the same if these questions are seriously attended to. It becomes increasingly evident that only a diffused model of religious leadership can resonate with people on the margins. Only such a model can acknowledge and appreciate people as subjects of their lived religion. This realization can lead to a rethinking on the ecclesiology that is being upheld today.

CONCLUSION

The traditional mode of marine fishing remains one of the most risk-ridden occupations and it has been sustaining a population along the seashore for generations. The present book was an attempt to explore the inner labyrinth of a maritime community in Kerala in order to grasp the secret of their amazing courage and inner strength to stay on in this field. The fisherfolk of the Thiruvananthapuram coast possess a remarkable spirit of bravery that allows them to venture into the roaring sea without fear. Danger and possible death do not deter them, and their resilience even after a disaster is incredible. What could be the source of this amazing courage, vitality, and resilience? This was the key question that prompted the present exploration.

The study kept its orientation based on three hypotheses. First, the fisherfolk have an organic relationship with the ocean and all marine life that enables them to see the ocean not as a place of danger but one of nourishment and life support. Two, the fisherfolk are a deeply religious people, firm in their faith convictions, and devoted to their ritual practices, and this religiosity sustains them before risk and danger. Three, the Catholic fisherfolk have incorporated elements from their neithal religious heritage to mold their own Catholicism to suit their life context. The nine chapters thematically organized the empirical data gathered from the Catholic fisherfolk along the Thiruvananthapuram coast in Kerala in two stages, first in 1995–1996, and later in 2018 soon after the Ockhi disaster. The book explored how the fisherfolk live their religion at the intersection between the sea and the land mass by articulating their worldview and surveying the way they mold their own Neithal Catholicism.

A closer examination of the lived religion of the fisherfolk revealed two distinct faith traditions actively at work. They had inherited an Indic/neithal religiosity that fashioned their outlook on life and work from ancient times. Along with that, they wholeheartedly welcomed the Iberian version of Catholicism that European missionaries brought along in the middle of the sixteenth century. What is remarkable is that they managed to accept the new without renouncing the old. More than that, they succeeded to incorporate the required elements of both to mold what could be called Neithal Catholicism. This was done to meet their religious aspirations so that they could draw strength from it in a risk-ridden occupation with courage and unfailing hope. The Iberian Catholicism provided them with a new identity and pride as part of a worldwide fellowship of believers as they lived in a caste-ridden society. At the same time, they continued to rely on a meaning frame that was entrenched in the neithal religiosity and worldview. The cyclone of 2018, the first one of such intensity in their lifetime, seemed to shatter their faith in a protective Mother Ocean, but their resilience in getting their life back into the waters proved their unfailing hope beyond any disaster. What is of interest to students of religion is the skilful way they crafted their Neithal Catholicism through a process of inter-cultural encounter and dialogue. Here, the fisherfolk themselves become architects of their religion and administrators of its rituals in a diffused mode. It is the religion they live, and it is their religion. It gave them strength to survive in an environment that is not always hospitable.

The religious journey of the fisherfolk has not been a totally smooth and harmonious one. They were aware of certain incongruities and have been struggling to deal with them. One area of clear difficulty was concerning the religious leadership of the community. They always had trouble with a centralized and all-encompassing mode of priesthood that was brought by the Iberian Catholicism. The same conflict and struggle seem to continue even today. The new generation coming from coastal communities realizes the need for promoting a diversified mode of leadership—social, cultural, and political—that is not dependent on a hegemonic religious leadership. The experience of being sidelined in a participatory form of democracy and failing to get a proper share of benefits forces them to radically review the leadership roles. Equally noteworthy is a vacuum in the area of cultural expressions that continued from the time of the mass conversion since the official church had frowned upon all traditional cultural forms as *pagan*. The vacuum is deeply felt by the new generation, but it will require concerted efforts to refill it.

The State of Kerala witnessed many reform movements in the nineteenth through twentieth centuries led by visionaries like Narayana Guru, Ayyankali, and Mannath Padmanabhan. These movements helped many communities to tread the path of emancipation and empowerment but left the coastal communities largely untouched. A new awareness about the dormant inner strength along with a new-found pride in the neithal heritage has encouraged the new generation to explore its cultural roots. Unfortunately, documented history is mostly limited to the post-conversion period, thanks to the missionaries. But their reports and letters portray a European viewpoint, and seldom do justice to the neithal culture and worldview. With the end of the Sangam period, the documented history of the coastal communities seems to have vanished for many centuries. This paucity of data remains a major hurdle for researchers, and so demands the application of new methods and tools of inquiry.

The experiences and expressions of marginal groups in a society rarely find mention in the chronicles of royal history and textbooks of academicians. The same logic would explain why Christian theological productions and doctrinal pronouncements tend to be silent about the faith life and practices of people like the fisherfolk. A serious study of the lived Catholicism of the coastal people becomes significant in such a context. The fisherfolk remind us that the seashore is no more the outer boundary of the church, nor are the coastal people marginal to the Christian fellowship. Instead, they show us that the ocean is the real center and human life on earth is amazingly dependent on the ocean and its mysteries. Human life itself is a moment between the blue sky and the vast ocean. This realization makes us humble but gives us courage and hope to look at the ocean and the horizon beyond it.

The lived religion of the coastal people opens up new avenues for religious studies and for the discipline of theology to explore. Specially significant are issues related to ecclesiology with its conventional notion of closed boundaries, meaning of sacraments within a broader frame of rituals, the role of other religious traditions in Christian soteriology, the motherly aspect in the conception of the divine, a review of the Christian virtues in the light of the *rasa* theory, and finally, a rethinking on the priestly role in terms of a diffused model of religious leadership. It is beyond the scope of this book to take up these for detailed discussion, but these surely indicate the way to move forward in order to enrich Christian theology and Christian life. My hope is that these findings will provide the building blocks for an authentic Asian Christian theology that may resonate with the aspirations of people of the land—and of the seashore.

GLOSSARY

[*Terms included in this glossary follow the Malayalam usage, except when Tamil or Sanskrit terms require their original form.*]

Amma	mother; a reverential term that is used for any person of the female gender, for mother goddesses, for Mother Ocean, etc.
Arayas	a Hindu fishing caste (also called Nuḷayan)
Asuras	demonic creatures in Hindu mythology
Ayurveda	an Indian system of medicine
Bhadrakāḷī	one of the village goddesses linked to Śakti worship
Bhagavathy	local name for Kāḷī or Bhadrakāḷī
bhāva	expression of characteristic emotional state or sentiment by bodily gestures or by words
cakram	mantric wheel that is used in *mantravādam* and as a weapon; weapon of Vishnu
chaviṭṭunāṭakam	a Christian form of dance drama that was popular in the coastal regions of central Kerala
cēluparachil	ritual evaluation of a fishing operation by the crew members
Chemmeen	name of a popular novel in Malayalam; it also means shrimps.
Cintāthira Māta	Our Lady of Sea Voyage; corrupt form of Sindhu-Yatra-Māta, patron of the church at Vizhinjam
sabhakal	confraternities or pious associations of lay persons in a Catholic parish; *sabha* (sing.)
Dēvī	a goddess, a female deity; divine feminine
Dhīvara	generic term, meaning fisherman, popularly used for Hindu fishing castes

eilamidal	collective singing of traditional occupational songs during fishing operations, especially shore-seine operations
eilass	amulet used in *mantravādam*; metal casket enclosing a *mantra* written on a metal foil or paper, worn by a person on the body or tied to an object in order to ensure good fortune or to ward off evil
JātiTalaivan	traditional caste head-man of the Paravas of the Coromandel Coast of Tamil Nadu
kadal	sea, ocean; also used for the breakers
Kadalamma	Mother Ocean; the ocean conceived as a deity
Kāḷī	one of the female deities usually associated with Śakti worship
kaṇiyam	the skill of computing astronomical configurations in a fishing context
kampēru	evil eye (also referred to as *kaṇṇupedal*)
kavaru	phenomenon of glow in the sea water, visible in the dark
Kenkādeviamma	reverential term used for Mother Ocean (a corrupt form of *Gangā-devī-amma*=mother goddess of the sea)
kudumbakkār	people belonging to the elite families, traditionally owners of craft and gear
kuthaka	tax levied by the church on the catch of fish
mantra, mantram	sacred formula or mystical verse addressed to a deity in religious ritual; a charm, spell, or incantation
mantravādam	the science of magic spells; the practice of mantric rituals
mantravādi	practitioner of *mantravādam*; an exorcist
Māta	mother, mother goddess; used for Virgin Mary in the Christian context
meen	fish; the same term is used for stars and planets
muni	a sage, saint, or ascetic; a category of spirits appearing with terrifying looks
nāvichi	*nāvier* woman
nāvier	non-fishing barber group in a fishing village

neithaltiṇai	the coastal tract, one of the five geographical and cultural zones mentioned in the Sangam literature
nercha	ritual offering to a deity
Nuḷayas	a Hindu fishing caste (also called Arayan)
Onam	harvest festival that is celebrated in Kerala in August-September
Padroado	patronage; a system of benefice granted by the Holy See to Portugal and Spain, in return for their role in the propagation of faith in new territories during the colonial period
Paravas	a fishing caste mainly engaged in pearl fishing along the Coromandel Coast; Parava–adj. (Also known as Parathavar or Bharathas)
pāru	under-water cleft in the sea that serves as breeding ground and shelter for fish; rich fishing ground
peiy	demonic spirits; spirit of the dead that is believed to roam around the earth and disturb the living
Pirusenti	President of a confraternity (*compiria sabha*) in a parish
piśāśu	the devil, demon, evil spirit
pulikiyar	the worker group in fishing, usually dependent on the *kudumbakkār*
Pradhāni	headman of the village; a member of the ruling elite
rasa theory	theory in Indian aesthetics that systematizes *rasas* and *bhāvas*
rudrabhāva	terrifying nature
Śakti	divine power or energy; female deity; the divine feminine
Sangam poetry	collection of the literary works of poets belonging to the ancient Tamil Academy or Sangam that existed between 500 BCE and 200 CE
śāntabhāva	benign nature
Santiyagappar	Santiago, or St. James the Great
śāpam	curse, spell
Saptamātṛkaḷ	Seven Mothers; a group of mother goddesses, seven in number

Śiva	a proto Indian god, later assimilated into the Vedic pantheon and integrated with Rudṛa-Śiva, finally becoming one of the *Trimurthis*
Tamiḷakam	name of the ancient Dravidian country, comprising present-day Tamil Nadu and Kerala, and also part of Karnataka.
Tantrism	the teaching of the *Tantras*, the religious texts dealing with the esoteric practices of some Hindu, Buddhist, and Jaina sects. In the present context, it refers mainly to *Śākta Tantras* that emphasize Śakti as the female personification of the creative energy of the divine.
Ūramma	protective mother goddess of a village
ūruvilakku	ostracism; system of community sanction that existed in fishing villages, now disappearing
vaidyan	practitioner of Indian system of medicine like Ayurveda or Siddha
Vairavan	a Dravidian deity, identified as a fierce form of Śiva, propitiated in māntric rituals (also called Bhairavan)
vēl	spear or lance; weapon of Murugan
Vēlan	Murugan, one who bears the *vēl* (spear)
veḷḷi	silver; generic name used for stars and planets
yantra	magical design; diagram of geometrical patterns used as mystical symbol in rituals

SELECTED BIBLIOGRAPHY

A. BOOKS AND ARTICLES IN ENGLISH

Abraham, C.M. *Fishworkers Movement in Kerala*. Mumbai: Institute for Community Organization Research, 1995.

Aerthayil, Mathew. *The Fishermen's Movement in Kerala*. Delhi: Indian Social Institute, 2000.

Agur, C.M. *Church History of Travancore*. New Delhi: Asian Education Services, 1990.

Amaladoss, Michael. *Becoming Indian: The Process of Inculturation*. Bangalore: Dharmaram Publications, 1992.

Antony, E.P. *The History of Latin Catholics in Kerala*. Ernakulam: I.S.Press, 1992.

Attwater, Donald. *A Dictionary of Mary*. London: Longmans, 1957.

Balasubramanian, C. *A Study of the Literature of the Cera Country* (Upto 11th Century). Madras: University of Madras, 1980.

Baum, Gregory. *The Social Imperative: Essays in the Critical Issues that Confront the Christian Churches*. New York: Paulist, 1979.

Bayly, Susan. *Saints, Goddesses and Kings: Muslims and Christians in South Indian Society, 1700–1900*. Cambridge: Cambridge University Press, 1989.

Bell, Catherine, *Ritual Theory, Ritual Practice*. New York: Oxford University Press, 1992.

Bharati, Agehananda. *The Tantric Tradition*. New York: Samuel Weiser Inc., 1975.

Bhattacharyya, N.N. *History of the Tantric Religion*. New Delhi: Manohar Publishers, 1982.

Brodrick, James. *Saint Francis Xavier*. London: Burns Oates, 1958.

Brown, Peter. *The Cult of the Saints: Its Rise and Function in Latin Christianity*. London: S.C.M. Press Ltd., 1981.

Chandran, J. Russell, ed. *Third World Theologies in Dialogue* (Essays in Memory of D.S. Amalorpavadass). EATWOT, 1991.

Chawdhri, L.R. *Secrets of Yantra, Mantra and Tantra.* New Delhi: Sterling Publishers, 1992.

Chiramel, Benny. *The Phenomenon of Social Exclusion of Fisherfolk Youth in relation to their Self-esteem, Aggression and Religious Identity.* Unpublished PhD Dissertation, University of Mumbai, 2014.

Clausen, Arne Martin. *Kerala Fishermen and the Indo-Norwegian Pilot Project*. London: Allen and Unwin, 1968.

Cohen, P. Antony. *The Symbolic Construction of Community.* London: Tavistock Publications, 1985.

Coleridge, Henry James. *The Life and Letters of St. Francis Xavier.* 2 Vols. London: Burns Oates, 1881.

Congar, Yves M.J. *A History of Theology.*Translated and edited by Hunter Guthrie. New York: Doubleday and Company, 1968.

Correia-Afonso, John. *Jesuit Letters and Indian History 1542-1773.* Bombay: Oxford University Press, 1969.

Da Silva, R.William, ed. *Discoveries, Missionary Expansion and Asian Cultures.* New Delhi: Concept Publishing Company, 1994.

de Souza, Teotonio R. *Goa to Me.* New Delhi: Concept Publishing Company, 1994.

De Vaux, Roland. *Ancient Israel: Its Life and Institutions.* Translated by John McHugh. London: Darton, Longman and Todd, 1986.

Directorate of Fisheries. *Marine Fisheries of Kerala at a Glance.* Trivandrum: Government of Kerala, 1991.

Dumont, Louis. *A South Indian Sub-caste: Social Organization and Religion of the Pramalai Kallar.* Delhi: Oxford University Press, 1986.

Economic Review 2016—Agriculture and Allied Sectors, Fisheries Development. Thiruvananthapuram: State Planning Board, 2016.

Ferroli, Domenico. *The Jesuits in Malabar.* 2 vols. Bangalore: The National Press, 1939–1951.

Firth, Raymond. *Malay Fishermen and their Peasant Economy.* London: R.K.P, 1971.

Firth, Raymond. *Human Types: An Introduction to Social Anthropology.* New York: Mentor Books, 1958.

Flannery, Austin, ed. *Vatican Council II: The Conciliar and Post-Conciliar Documents.* Bombay: St. Paul Publication, 1988.

Gennep, Arnold Van. *The Rites of Passage.* Translated by M.B. Vizedom and G.L. Caffee. Chicago: The University of Chicago Press, 1960.

Gill, Robin, ed. *Theology and Sociology: A Reader.* London: Geoffrey Chapman, 1987.

Goffman, Erving. *The Presentation of Self in Everyday Life*. New York: Doubleday, 1959.

Government of Kerala. *Kerala State Gazetteer,* Vol.II, Part I. Thiruvananthapuram: 1986.

Government of Kerala. *Kerala State Gazetteer,* Vol.III. Thiruvananthapuram: 1989.

Government of Kerala. *Vizhinjam Fishery Harbour—Project Report*. Development Department, 1979.

Greeley, Andrew M. *The Mary Myth: On the Femininity of God*. New York: Crossroad Book, 1977.

Gulati, Leela. *Fisherwomen on the Kerala Coast*. Geneva: International Labour Office, 1984.

Hammond, E. Philip. *The Sacred in a Secular Age: Towards Revision in the Scientific Study of Religion*. Berkeley: University of California Press, 1985.

Hemingway, Ernest. *The Old Man and the Sea*. London: Penguin Books, 1952.

Hill, Michael. *A Sociology of Religion*. London: Heinemann, 1980.

Houtart, F. and G. Lamercinier. *Genesis and Institutionalization of the Indian Catholicism*. Louvain: C.R.S.R., 1981.

Houtart, Francis, and Nalini Nayak. *Kerala Fishermen: Culture and Social Organization*. Trivandrum: PCO Centre, 1988.

Iyengar, P.T. Srinivas. *History of the Tamils: From the Earliest Times to 600 A.D.* New Delhi: Asian Educational Services, 1983.

Iyer, L.K. Ananthakrishna. *The Travancore Tribes and Castes*. Trivandrum: Government Press, 1937–1939.

Jeyechandrun, A.V. *Sakti*. Madurai: The Temple Art Museum Series No.1, 1966.

John, K.J. *Christian Heritage of Kerala,* Grand Chevalier L.M. Pylee Felicitation Volume. Cochin: 1981.

Kakar, Sudhir. *Shamans, Mystics and Doctors: A Psychological Inquiry into India and its Healing Traditions*. New York: Alfred A. Knoff, 1982.

Kalawar, A.G. *Report of the Expert Committee on Marine Fisheries in Kerala*. Trivandrum: Government of Kerala, 1985.

Kariyil, Antony. *Church and Society in Kerala: A Sociological Study*. New Delhi: Inter-cultural Publications, 1995.

Kattikaran, Kurien. "Determinants of the Meaning System Among the Artisanal Fish Workers of Quilon Town". Unpublished MSW dissertation, Trivandrum: Loyola College of Social Sciences, 1983.

Kinsley, David. *Hindu Goddesses: Visions of the Divine Feminine in the Hindu Religious Tradition*. Delhi: Motilal Banarsidass, 1987.

KIRTADS. *The Hindu Fisherfolk of Kerala: Why Were They Not Classified as Scheduled Caste?* Kozhikode: Department of Kirtads, 1983.

Klausen, Arne Martin. *Kerala Fishermen and the Indo-Norwegian Pilot Project*. London: Allen and Unwin, 1968.

Koilparampil, George. *Caste in the Catholic Community in Kerala*. Cochin: St. Teresa's College, 1982.

Kosambi, D.D. *Myth and Reality: Studies in the Formation of Indian Culture*. Bombay: Popular Prakashan, 1962.

Krishna, G.R. *Caste and Tribes of Fishermen*. New Delhi: Discovery Publishing House, 1990.

Kumar, Ajit, "An Appraisal of Vilinjam's History and Archeology in the Wake of Recent Explorations." Pages 72–80 in *Kerala Spectrum–Aspects of Cultural History*. Edited by Charles Dias. Cochin: The Indo-Portuguese Cultural Institute, 2006.

Kurien, John. "The Kerala Model: Its Central Tendency and the Outlier." Paper presented at the International Congress on Kerala Studies, Trivandrum, 1994.

Kurien, John. "Towards an Understanding of the Fish Economy of Kerala State." Trivandrum: C.D.S. Working Paper No. 68, 1978.

Lamercinier, Genevieve. *Religion and Ideology in Kerala*. Louvain: C.R.S.R., 1983.

Luckmann, Thomas. *The Invisible Religion*. London: The Mac Millan Co., 1967.

Luzbetak, Louis J. *The Church and Cultures: New Perspectivess in Missiological Anthropology*. New York: Orbis, 1988.

Mahapatra, Sitakant, ed. *The Realm of the Sacred: Verbal Symbolism and Ritual Structures*. Calcutta: Oxford University Press, 1992.

Mathur, P.R.G. *The Mappila Fisherfolk of Kerala: A Study in Inter-relationship between Habitat, Technology, Economy, Society and Culture*. Trivandrum: Kerala Historical Society, 1977.

McGuire, Meredith. *Religion: The Social Context*. California: Wadsworth Publishing Company, 1981.

Menon, P. Shungoonny. *History of Travancore*. Government Press, 1983.

Merton, K. Robert. *Social Theory and Social Structure* (1968 enlarged edition). New Delhi: Amerind Publishing, 1972.

Metford, J.C.J. *Dictionary of Christian Lore and Legends*. London: Thames and Hudson, 1983.

Ministry of Agriculture. *Marine Fisheries Census 2010, India, Part I*. Kochi: Governement of India and CMFRI, 2010.

Ministry of Agriculture. *Marine Fisheries Census 2010,* Part II, 6, Kerala. Kochi: Government of India and CMFRI, 2010.

Morais, George Mark. *A History of Christianity in India: From Early Times to St. Francis Xavier, A.D. 52-1542*. Bombay: Manaktalas, 1964.

Mundadan, A. Mathias. *History of Christianity in India, Vol. I: From the Beginning upto the Middle of the 16th Century*. Bangalore: Theological Publications in India, 1984.

Murickan, J. *Religion and Power Structure in Rural India*. Jaipur: Rawat Publications, 1991.

Murickan, Jose. *Storm on the Seashore: The Fishermen's Struggle in Kerala*. New Delhi: Indian Social Institute, 1987.

Nagam Aiya, V. *The Travancore State Manual*, 3 Vols. New Delhi: Asian Educational Services, 1989 (1906).

Nagaswami, R. *Tantric Cult of South India*. Delhi: Agam Kala Prakashan, 1982.

Nair, N. Balakrishnan. *Report of the Expert committee on Ban on Monsoon Trawling in Kerala*. Trivandrum: Government of Kerala, 1991.

Nair, N. Balakrishnan. *Report of the Expert Committee on Marine Fishery Resources Management in Kerala*. Trivandrum: Government of Kerala, 1989.

Nayak, Nalini. *Continuity and Change in Artisanal Fishing Communities: A Study Report*. Trivandrum: P.C.O. and SIFFS, 1993.

Neuner J. and J. Dupuis, eds. *The Christian Faith: In the Doctrinal Documents of the Catholic Faith*. 5th ed. Bangalore: Theological Publications in India, 1992.

O'Dea, Thomas and Janet O'Dea Aviad. *The Sociology of Religion*. 2nd ed. New Jersey: Prentice Hall, 1983.

O'Flaherty, Wendy Doniger. *The Origins of Evil in Hindu Mythology*. Berkeley: University of California Press, 1980 (1976).

Pals, Daniel L. *Seven Theories of Religion*. New York: Oxford University Press, 1996.

Pandian, J. *Caste, Nationalism and Ethnicity*. London: Sangam Books, 1987.

Pathil, Kuncheria, ed. *Religious Pluralism: An Indian Christian Perspective*. Delhi: ISPCK, 1991.

Paul, Babu. *Report of the Committee to Study the Need for Conservation of Marine Fishery Resources during Certain Seasons of the Year and Allied Matters*. Trivandrum: Government of Kerala, 1982.

Pieris, Aloysius, "Prophetic Humour and the Exposure of Demons: Christian Hope in the Light of a Buddhist Exorcism." *Vidyajyoti Journal of Theological Reflection* 60, no, 5 (May 1996): 311–322.

Pillai, J.M. Somasundaram. *Two Thousand Years of Tamil Literature: An Anthology with Studies and Translations.* Madras: 1959.

Platteau, Jean-Philippe and Jose Murickan. *Technology, Credit and Indebtedness in Marine Fishing: A Case Study of Three Fishing Villages in South Kerala.* Delhi: Hindustan Publishing, 1985.

Pushparajan, M. *Ampa: Essays on Fishermen's Folk Songs.* Jaffna: Alai Literary Circle, 1976.

Puthenangady, P. *Popular Devotions in India.* Bangalore: NBCLC, 1986.

Ram, Kalpana. *Mukkuvar Women: Gender, Hegemony and Capitalist Transformation in a South Indian Fishing Community.* New Delhi: Kali for Women, 1991.

Rangacharya, Adya. *Introduction to Bharata's Natya-Sastra.* Bombay: Popular Prakasan, 1966.

Rayan, Samuel. "The How of Third World Theologies." Edited by J. Russell Chandran. EATWOT: *Third World Theologies in Dialogue*, 1991.

Rayan, Samuel. *The Anger of God.* Bombay: BUILD, 1981.

Reddy, P.G. and A. Chellaperumal. "Religion Among the Irulas of Coimbatore District, Tamilnadu." Pages 76–86 in *Religion and Society in South India.* Edited by N.S. Reddy, V. Sudarsen, G.P. Reddy, and M. Suryanarayana. Delhi: B.R. Publishing Corporation, 1987.

Robinson, Rowena. "The Cross: Contestation and Transformation of a Religious Symbol in Southern Goa." *Economic and Political Weekly* 29, no. 3 (January 15), 1994: 94–98.

Roche, A. Patrick. *Fishermen of the Coromandel: The Social Study of the Paravas of the Coromandel.* New Delhi: Manohar, 1984.

Roche, E. Louis. "The Latin Catholic Community in Kerala, 1498–1653," PhD dissertation, University of Kerala, 1989.

Sachau, C. Edward, ed. *Alberuni's India.* Delhi: S. Chand and Co., 1964.

Sahay, N. Keshari. *Christianity and Cultural Change in India.* New Delhi: Inter-India Publications, 1986.

Schreiter, J. Robert. *Constructing Local Theologies.* New York: Orbis, 1985.

Schurhammer, George. *Francis Xavier: His Life, His Times.* Vol. II, India, 1541–1545. Translated by M. Joseph Costelloe S.J. Rome: The Jesuit Historical Institute, 1977.

Sharpe, J. Eric. *Understanding Religion.* London: Duckwoth, 1988.

Shorter, Aylward. *Jesus and the Witch Doctor: An Approach to Healing and Wholeness.* London: Geoffrey Chapman, 1985.

Stutley, Margaret and James Stutley. *A Dictionary of Hinduism: Its Mythology, Folklore and Development 1500 B.C-A.D.1500*. Bombay: Allied Publishers, 1977.

Subramaniam, V., ed. *Mother Goddess and Other Goddesses*. Delhi: Ajanta Publications, 1993.

Subramanian, N. *Sangam Polity: Administrative and Social life of the Sangam Tamils*. Madurai: Ennes Publications, 1980.

Tanaka, Masakazu. *Patrons, Devotees and Goddesses: Ritual and Power among the Tamil Fishermen of Sri Lanka*. New Delhi: Manohar, 1997.

Teasdale, Wayne. *Toward A Christian Vedanta: The Encounter of Hinduism and Christianity According to Bede Griffiths*. Bangalore: Asian Trading Corporation, 1987.

Thekkedath, Joseph. *History of Christianity in Kerala*, Vol. II. Catholic History Association, India. 1982.

Thomas P. *Christians and Christianity in India and Pakistan*. London: George Allen and Unwin Ltd., 1954.

Thurston, Edgar and K. Rangachari. *Castes and Tribes of Southern India*, Vol.5. New Delhi: Asian Educational Services, 1987.

Tillich, Paul. *Systematic Theology*. Chicago: The University of Chicago Press, 1971.

Turner, Roy, ed. *Ethnomethodology*. Middlesex: Penguin, 1974.

Turner, Victor. *The Forest of Symbols: Aspects of Ndembu Ritual*. London: Cornell University Press, 1977.

Turner, Victor. *The Ritual Process: Structure and Anti-Structure*. Middlesex: Penguin Books, 1972.

Udayabhanu, K.M. "The Dheevaras of Kerala: A Historical Perspective." *Journal of Kerala Studies* 14 (1987): 197–202.

Velacherry, K.J. Joseph. "A Study of a Fishing Community in Kerala." Unpublished PhD dissertation, University of Delhi, 1982.

Velupillai. *The Travancore State Manual*, 4 Volumes. Trivandrum: 1931.

Vimalakumari, T.K. *Infant Mortality Among Fishermen*. New Delhi: Discovery Publishing House, 1991.

Von Rad, Gerhard. *The Problem of the Hexateuch and Other Essays*. London: SCM Press, 1984, chs.14 and 15.

Ward, Benjamin S. and Peter E. Conner.*Memoir of the Survey of the Travancore and Cochin States*, Vol. I., rev. ed. Thiruvananthapuram: Government of Kerala, 1994.

Wilfred, Felix. *Sunset in the East: Asian Challenges and Christian Involvement*. Chennai: University of Madras, 1991.

Wilfred, Felix. "Towards a Subaltern Hermeneutics." *Jeevadhara* 26, no. 151 (January, 1996): 56–59.

Wilfred, Felix. *From the Dusty Soil: Contextual Reinterpretation of Christianity*. Madras: University of Madras, 1995.

Xavier, Francis. *The Letters and Instructions of Francis Xavier*. Translated by M. Joseph Costelloe. Anand: Gujarat Sahitya Prakash, 1993.

Yinger, Mitten. *The Scientific Study of Religion*. New York: Mac Millan, 1970.

Zacharia, Scaria, ed. *The Acts and Decrees of the Synod of Diamper, A.D.1599*. Hosanna Mount: Indian Institute of Christian Studies, 1994.

Zvelebil, Kamil V. *Tiru Murugan*. Madras: International Institute of Tamil Studies, 1981.

B. BOOKS AND ARTICLES IN MALAYALAM

Desabandhu Publications. *Kerala Vijñana Kosam*. Thirvanathapuram: Desabandhu Publications, 1988.

George, K.M. *Kerala Charitram*, Vol. 2. Ernakulam: Kerala History Association, 1974: 599–601.

Ibrahimkunju, A.P. *Marthanda Varma-Adhunika Thiruvithamkurinte Udayam*. Thiruvananthapuram: Government of Kerala, 1990.

Iyer, S.Sankara. "Buddha-Jaina Mathangal". In *Kerala Charitram*, Vol.2, Ernakulam: Kerala History Association, 1974: 327–370.

Iyer, Ulloor Parameswara. *Kerala Sāhitya Charitram*, Vol. 3. Thiruvananthapuram: Kerala University, 1972.

Jayakumar and Johnson Roche. "Sarkar Padhathikalil Kurungunna Vizhinjam." *Theerasabdam*, 7/73 (July 1985): 23–27.

Karim, C.K. *Francis Bukkanante Keralam* (A Journey from Madras through the Countries of Mysore, Canara and Malabar, Vol.2). Thiruvananthapuram: Kerala Bhasha Institute, 1981.

Kunjanpilla, Ilamkulam. *Kerala Charitrathile Iruladañja Edukal*. Kottayam: National Book Stall, 1983.

Latin Christian, Weekly Newspaper, published from Valiathura, issues from 16 August 1934–16 May 1935.

Nair, N. Balakrishnan. *Kadal Oru Adbhudam*. Trivandrum: State Institute of Languages, 1994.

Oru Samarakatha. Thiruvananthapuram: Kerala Swathantra Malsya Thozhilali Federation, 1988.

Panipilla, Robert. *Kadalarivukalum Neravubhavangalum*. Kottayam: Current Books, 2015.

Pilla, Thakazhi, Sivasankara. *Chemmeen*. Kottayam: D.C. Books, 1995 (1956).

Raphi, Sebeena. *Chaviṭṭunāṭakam*. Kottayam: National Book Stall, 1980 (1964).

Thankappanachary, T.R. *Kadal Malsyolpadanam Keralathil-Oru Avalokanam*. Thiruvananthapuram: P.C.O., 1986.

Thikkotiyan. *Chuvanna Kadal*. Trichur: Mangalodayam, 1963.

Visalakshi, N.R. "Arayanmar". In *Kerala Charitram*, Vol. 1. Ernakulam: Kerala History Association, 1973: 993–1011.

INDEX

G

I

J

K

L

M

N

P

R

S

T

U

V

W